D0849618

ORGANIC SYNTHESES VIA BORANES

HERBERT C. BROWN

Department of Chemistry
Purdue University
West Lafayette, Indiana

ORGANIC SYNTHESES VIA BORANES

with techniques by

Gary W. Kramer
Alan B. Levy
M. Mark Midland

Department of Chemistry,
Purdue University
West Lafayette, Indiana

A WILEY-INTERSCIENCE PUBLICATION

JOHN WILEY & SONS
New York • Chichester • Brisbane • Toronto

Copyright © 1975, by John Wiley & Sons, Inc.

All rights reserved. Published simultaneously in Canada.

Reproduction or translation of any part of this work beyond that permitted by Sections 107 or 108 of the 1976 United States Copyright Act without the permission of the copyright owner is unlawful. Requests for permission or further information should be addressed to the Permissions Department, John Wiley & Sons, Inc.

Library of Congress Cataloging in Publication Data

Brown, Herbert Charles, 1912-
Organic syntheses via boranes.

"A Wiley-Interscience publication."
Includes bibliographical references and index.
1. Organoboron compounds. 2. Chemistry, Organic--Synthesis. I. Title.

QD412.B1B77 1975 547'.05'671 74-20520
ISBN 0-471-11280-1

Printed in the United States of America

10 9 8 7 6 5 4

To My Research Group -
their dedication and enthusiasm
enhanced the pleasure of the research

ACKNOWLEDGMENT

The author is indebted to Gary W. Kramer, M. Mark Midland, and Alan B. Levy, who originally suggested the desirability of including a section on techniques and then accepted the responsibility for preparing that section. Gary W. Kramer was primarily responsible for the diagrams, with the excellent final drawings made by Leland E. Schatzley, Jr., and Barbara L. Schatzley (LeBar Graphics). Finally, Annette Wortman did her usual superlative job in typing the manuscript.

PREFACE

The remarkably facile addition of diborane in ether solvents to alkenes and alkynes was discovered in 1956. For the next decade a major portion of the research effort of my students and associates was devoted to the study of this fascinating new reaction, hydroboration.

This reaction made the organoboranes readily available. However, organoboranes had not been of special interest in the past, and they had received relatively little research attention. It appeared desirable, therefore, to undertake a program to explore these derivatives in more detail. Consequently, the emphasis of our program was shifted from the study of hydroboration to a study of the chemistry of organoboranes.

This proved to be an extraordinarily rich new area. There resulted a veritable explosion of new chemistry whose full potentiality we can only dimly visualize.

The problem is how to transmit the knowledge in the area to the chemists who must use the chemistry if its full potential is to be achieved.

Before us lies the utilization of these methods for the synthesis of complex molecules, such as natural products and pharmaceuticals. Before us lies the exploration of the applicability of this chemistry for the synthesis of fine chemicals. Before us lies the exploration of the utility of this chemistry in the petrochemical area.

But this is only the beginning. Still to be explored are also the reaction mechanisms involved in the remarkably clean reactions of the organoboranes. The spectroscopy of organoboranes is in its infancy. Structural effects have yet to be examined systematically.

Clearly it will require another generation of chemists to fully explore this new continent. However, before the new generation of students can be taught the new chemistry, their teachers must learn it. It is clear that many chemists have hesitated to utilize these methods, and many instructors have hesitated to introduce them in their courses because of their inexperience in handling and working with hydroboration and organoboranes.

It is my hope that this book will serve to acquaint working chemists, teachers, and students with the chemistry and techniques of organoborane chemistry. A second effort to overcome this hurdle has been to persuade Dr. Alfred R. Bader of the Aldrich Chemical Company to set up a subsidiary, Aldrich-Boranes, Inc., to make readily available the basic chemicals and intermediates and certain specialized pieces of apparatus to facilitate application of these new methods by chemists. It is expected that both developments will contribute to surmounting the barrier.

A new continent has been discovered—it requires settlers to develop its riches to contribute to mankind.

HERBERT C. BROWN

West Lafayette, Indiana
August 1974

CONTENTS

FIGURES

ABBREVIATIONS

Ac:	acetyl
Ar:	aryl
9-BBN:	9-borabicyclo[3.3.1]nonane, BH
BMS:	borane-methyl sulfide complex
Bu:	butyl
d^{25}:	density at 25°
DCME:	α,α-dichloromethyl methyl ether
DG:	diglyme $[(CH_3OCH_2CH_2)_2O]$
DME:	1,2-dimethoxyethane
DMF:	dimethylformamide
DMSO:	dimethyl sulfoxide
Δ:	heat
EE:	diethyl ether
Et:	ethyl
HB:	hydroboration
IPC:	isopinocampheyl
IR:	infrared
LAH:	lithium aluminum hydride
Me:	methyl
NBS:	*N*-bromosuccinimide
[O]:	alkaline hydrogen peroxide oxidation, chromic acid oxidation
Ph:	phenyl
R:	alkyl
Sia:	Siamyl, 3-methyl-2-butyl
Thexyl:	2,3-dimethyl-2-butyl
THF:	tetrahydrofuran
VPC:	vapor phase chromatography
F:	Figure
N:	Note, **P2.4:N3** refers to Note 3 in Preparation 2.4.
P:	Preparation
S:	Survey
T:	Table

HERBERT C. BROWN

1

HYDROBORATION WITH BORANE: SURVEY

The facile addition of the boron-hydrogen bond to carbon-carbon multiple bonds of unsaturated organic derivatives makes the corresponding organoboranes readily available for application in organic synthesis.[1-3]

$$\text{>C=C<} + \text{H–B<} \rightarrow \text{H–}\overset{|}{\underset{|}{\text{C}}}\text{–}\overset{|}{\underset{|}{\text{C}}}\text{–B<}$$

$$\text{–C≡C–} + \text{H–B<} \rightarrow \underset{\text{H}}{\overset{|}{\text{C}}}\text{=}\underset{\text{B}<}{\overset{|}{\text{C}}}$$

One of the many interesting reactions that such organoboranes undergo (**S5** and **S7**) is the rapid and essentially quantitative oxidation with alkaline hydrogen peroxide.

$$\text{H–}\overset{|}{\underset{|}{\text{C}}}\text{–}\overset{|}{\underset{|}{\text{C}}}\text{–B<} + H_2O_2 \xrightarrow{OH^-} \text{H–}\overset{|}{\underset{|}{\text{C}}}\text{–}\overset{|}{\underset{|}{\text{C}}}\text{–OH} + \text{HOB<}$$

$$\underset{\text{H}}{\overset{|}{\text{C}}}\text{=}\underset{\text{B}<}{\overset{|}{\text{C}}} + H_2O_2 \xrightarrow{OH^-} \underset{\text{H}}{\overset{|}{\text{C}}}\text{–}\overset{|}{\text{C}}\text{=O} + \text{HOB<}$$

[1] H. C. Brown, *Hydroboration,* W. A. Benjamin, New York, 1962.
[2] H. C. Brown, *Boranes in Organic Chemistry,* Cornell University Press, Ithaca, N.Y., 1972.
[3] G. M. L. Cragg, *Organoboranes in Organic Synthesis,* Marcel Dekker, New York, 1973.

Consequently, hydroboration-oxidation provides a valuable procedure for achieving the hydration of carbon-carbon multiple bonds with high regio- and stereospecificities.[4]

1.1. HYDROBORATION PROCEDURES

Sodium borohydride is readily soluble in diglyme at 25°. It was originally observed that the addition of anhydrous aluminum chloride to such solutions not only enhanced the reducing power of the borohydride, but also achieved the hydroboration of olefins present in the reaction mixture.[5]

$$9\,RCH{=}CH_2 + 3\,NaBH_4 + AlCl_3 \xrightarrow{DG} 3(RCH_2CH_2)_3B + AlH_3 + 3\,NaCl$$

Replacement of the aluminum chloride by boron trifluoride made possible a more effective utilization of the hydride.

$$12\,RCH{=}CH_2 + 3\,NaBH_4 + 4\,BF_3 \xrightarrow{DG} 4(RCH_2CH_2)_3B + 3\,NaBF_4$$

Since sodium fluoborate is also readily soluble in the solvent, this reaction provides clear solutions of the organoborane for further operations (**P2.1**).

Sodium borohydride is only slightly soluble in tetrahydrofuran. Yet the reaction proceeds satisfactorily in that solvent.

$$12\,RCH{=}CH_2 + 3\,NaBH_4 + 4\,BF_3{:}OEt_2 \xrightarrow{THF} 4(RCH_2CH_2)_3B + 3\,NaBF_4\downarrow + 4\,Et_2O$$

The sodium fluoborate is insoluble in this solvent and precipitates. However, in most cases it is inert and need not be removed for subsequent operations (**P2.2**).

Diborane is highly soluble in tetrahydrofuran, existing in such solutions as the addition compound, $H_3B{:}THF$.[6] Such solutions are readily prepared (**P2.3**), and are now commercially available (borane-tetrahydrofuran complex[7]). In our work, we have found the treatment of the unsaturated organic compound with the calculated quantity of a standardized borane-THF solution to be the most convenient route to the corresponding organoborane (**P2.4, P2.5**).

[4] G. Zweifel and H. C. Brown, *Org. React.*, **13**, 1 (1963).
[5] H. C. Brown and B. C. Subba Rao, *J. Amer. Chem. Soc.*, **78**, 2582 (1956); **81**, 6423 (1959).
[6] B. Rice, J. A. Livasy, and G. W. Schaeffer, *J. Amer. Chem. Soc.*, **77**, 2750 (1950).
[7] Produced by Aldrich-Boranes, Inc., a subsidiary of Aldrich Chemical Company, Inc.

$$3\,RCH{=}CH_2 + H_3B{:}THF \xrightarrow{THF} (RCH_2CH_2)_3B + THF$$

This procedure provides the organoborane in a known quantity in a volatile, easily removed solvent, in the absence of inorganic salts or other undesirable side products.

The complex of borane with methyl sulfide is more stable than that with THF.[8] Consequently, it is possible to prepare and handle the pure 1:1 addition compound, $H_3B{:}S(CH_3)_2$. The borane-methyl sulfide complex[7] is active for hydroboration.[8] Not only does it provide the hydroborating agent in a highly concentrated form, but it makes possible hydroborations in a wide variety of solvents, such as ethyl ether and hexane[9] (**P2.6, P2.8**).

$$3\,RCH{=}CH_2 + H_3B{:}S(CH_3)_2 \xrightarrow{EE} (RCH_2CH_2)_3B + (CH_3)_2S$$

A possible disadvantage is the presence of methyl sulfide in the reaction mixture. Fortunately, it does not interfere with the hydrogen peroxide oxidation (**P2.6**), and its volatility (bp 35°) makes it easy to remove from the organoboranes (**P2.8**).

A wide variety of amine-boranes is also available[7] and certain of these are applicable for hydroborations.[10] However, they have not yet been explored in any detail. Most of them suffer from the disadvantage that they require somewhat elevated temperatures to achieve reaction. Some of the organoboranes are labile and should be protected from such temperatures.

Finally, it should be mentioned that a wide variety of both simple and complex metal hydrides, solvents, and acids can be utilized to achieve hydroboration.[11] However, these modifications do not appear to offer significant advantages over the four simple procedures (**P2.1**, **P2.2**, **P2.4**, and **P2.6**) utilized in the representative syntheses described here.

1.2. SCOPE AND STOICHIOMETRY

Hydroboration of the great majority of simple olefins proceeds simply and rapidly to the corresponding organoboranes. Alkenes containing two, three, or

[8] L. M. Braum, R. A. Braum, H. R. Crissman, M. Opperman, and R. M. Adams, *J. Org. Chem.*, **36**, 3888 (1971).
[9] C. F. Lane, *J. Org. Chem.*, 39, 1437 (1974).
[10] R. Koster, *Angew. Chem.*, 68, 684 (1957).
[11] H. C. Brown, K. J. Murray, L. J. Murray, J. A. Snover, and G. Zweifel, *J. Amer. Chem. Soc.*, **82**, 4233 (1960).

four alkyl substituents on the double bond readily undergo hydroboration. Cyclic and bicyclic olefins, such as 1,2-dimethylcyclopentene and α-pinene, readily react. Aryl groups on the double bond are accommodated, as in 1-phenylcyclohexene, 1,1-diphenylethylene, *trans*-stilbene, and triphenylethylene.[12] Finally, a number of steroids with highly hindered double bonds have been demonstrated to react.[13]

The great majority of olefins undergo complete reaction to form the corresponding trialkylborane (**P2.7**). However, in the case of more hindered

$$3\ (H_3C)HC{=}CH(CH_3) + BH_3 \longrightarrow (H_3C)HC(H){-}CH(CH_3){-})_3B$$

alkenes, such as trimethylethylene and tetramethylethylene, the reaction proceeds rapidly only to the dialkylborane (disiamylborane, **P2.9**) or the monoalkylborane stage (thexylborane, **P2.10**).

$$2\ (H_3C)_2C{=}CH(CH_3) + BH_3 \longrightarrow (H{-}C(CH_3)_2{-}CH(CH_3){-})_2BH$$

$$(H_3C)_2C{=}C(CH_3)_2 + BH_3 \longrightarrow H{-}C(CH_3)_2{-}C(CH_3)_2{-}BH_2$$

In such cases it is often possible to force the reaction to proceed to a subsequent stage, such as trisiamylborane or dithexylborane,[14] by utilizing relatively concentrated solutions and long reaction times. In the case of relatively stable alkyl moieties, higher temperatures can be used to drive the reaction to completion (**P6.12**).

On rare occasions it has been noted that molecules containing double bonds deeply buried within the structure, such as 5-α-cholest-8(14)-en-3-β-ol, fail to undergo hydroboration.[13]

[12] H. C. Brown and B. C. Subba Rao, *J. Amer. Chem. Soc.*, **81**, 6428 (1959).
[13] M. Nussim, Y. Mazur, and F. Sondheimer, *J. Org. Chem.*, **29**, 1120 (1964).
[14] E. Negishi, J. -J. Katz, and H. C. Brown, *J. Amer. Chem. Soc.*, **94**, 4025 (1972).

However, it is not clear that such cases have yet been subjected to forcing conditions.

1.3. DIRECTIVE EFFECTS

Simple 1-alkenes, such as 1-hexene, undergo hydroboration to place 94% of the boron on the terminal position with 6% at the 2-position. This distribution is not influenced significantly by branching of the alkyl group.[15]

$CH_3(CH_2)_3CH{=}CH_2$ — 6% ↑ (C-2), 94% ↑ (C-1)

$(H_3C)_3C{-}CH{=}CH_2$ — 6% ↑ (C-2), 94% ↑ (C-1)

On the other hand, an aryl group, such as that in styrene, causes increased placement of the boron atom on the nonterminal position, and this distribution is markedly altered by substituents in the ring.[16]

$p\text{-}CH_3O{-}C_6H_4{-}CH{=}CH_2$: 7%

$C_6H_5{-}CH{=}CH_2$ (H): 19%

$p\text{-}Cl{-}C_6H_4{-}CH{=}CH_2$: 27%

An alkyl substituent in the 2-position enhances attachment of the boron to the terminal carbon atom.[15]

[15] H. C. Brown and G. Zweifel, *J. Amer. Chem. Soc.*, 82, 4708 (1960).
[16] H. C. Brown and R. L. Sharp, *J. Amer. Chem. Soc.*, 88, 5851 (1966).

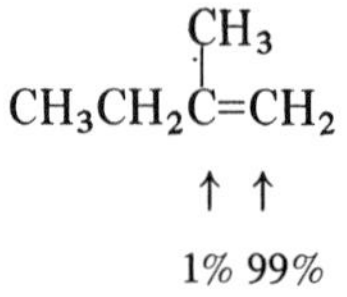

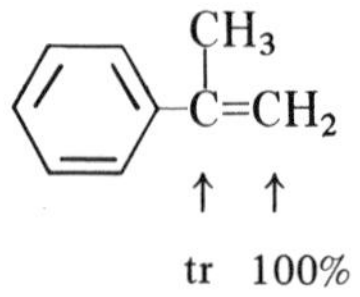

$$\underset{\uparrow\ \ \uparrow}{CH_3CH_2\overset{CH_3}{\overset{|}{C}}{=}CH_2} \qquad 1\%\ 99\%$$

$$C_6H_5{-}\overset{CH_3}{\overset{|}{C}}{=}CH_2 \qquad \text{tr}\ \ 100\%$$

A similar preference for the less substituted position is exhibited in internal olefins.

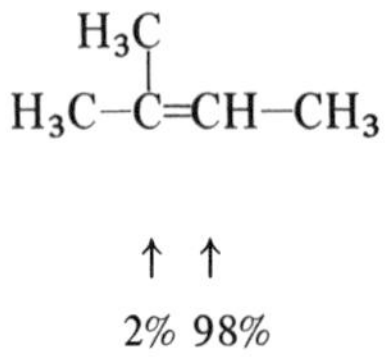

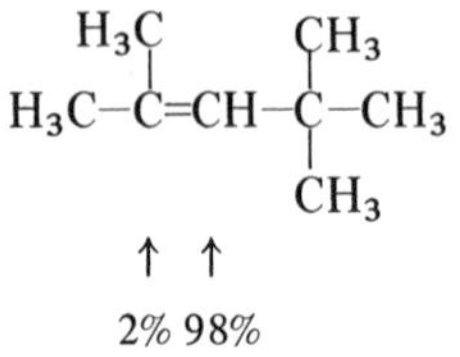

$$H_3C{-}\overset{H_3C}{\overset{|}{C}}{=}CH{-}CH_3 \qquad 2\%\ 98\%$$

$$H_3C{-}\overset{H_3C}{\overset{|}{C}}{=}CH{-}\overset{CH_3}{\underset{CH_3}{\overset{|}{\underset{|}{C}}}}{-}CH_3 \qquad 2\%\ 98\%$$

On the other hand, there is no significant discrimination between the two positions of an internal olefin containing alkyl groups of markedly different steric requirements.

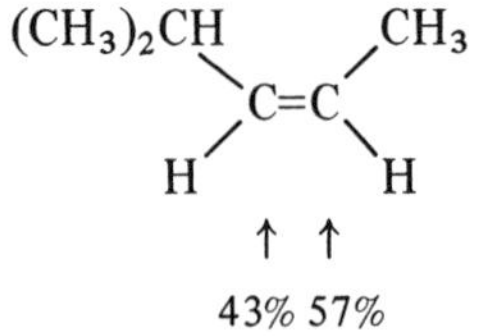

$$(CH_3)_2CH(H)C{=}C(H)CH_3 \qquad 43\%\ 57\%$$

(However, as discussed in **S3**, it is possible to achieve selective hydroboration of such olefins by use of dialkylboranes as selective hydroborating agents.)

It appears that the hydroboration reaction involves a simple four-center transition state, with the direction of addition controlled primarily by the polarization of the boron-hydrogen bond, $>B^{\delta+}{-}H^{\delta-}$.

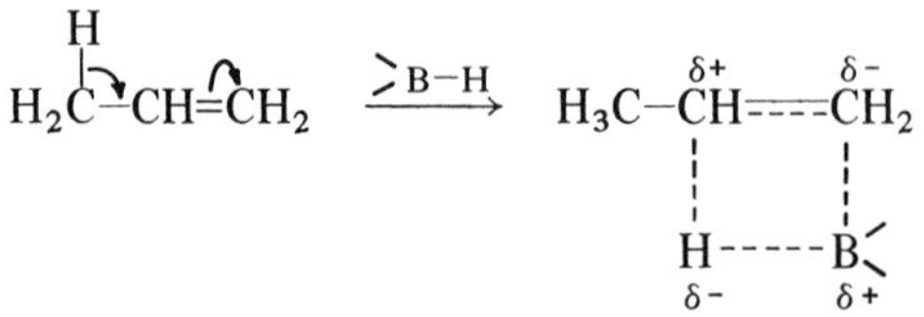

$$H_2\overset{H}{\overset{|}{C}}{-}CH{=}CH_2 \xrightarrow{>B-H} H_3C{-}\overset{\delta+}{CH}{\cdots}\overset{\delta-}{CH_2} \quad (\overset{\delta-}{H}{\cdots}\overset{\delta+}{B}<)$$

1.4. STEREOCHEMISTRY OF HYDROBORATION

The hydroboration of cyclic olefins, such as 1-methylcyclopentene and

1-methylcyclohexene, followed by oxidation with alkaline hydrogen peroxide, results in the formation of pure *trans*-2-methylcyclopentanol and *trans*-2-methylcyclohexanol **(P2.4)**.[17] Since the hydrogen peroxide oxidation evidently proceeds with retention of configuration, the hydroboration reaction must involve a *cis* addition of the hydrogen-boron bond to the double bond.

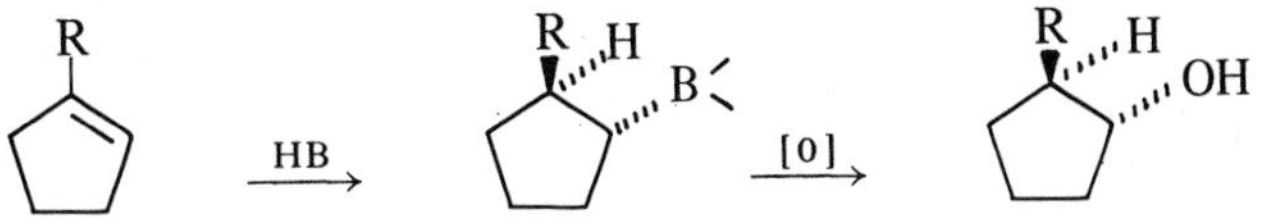

R HB R H B [O] R H OH

This *cis* hydration has been utilized to achieve a convenient synthesis of diastereomeric alcohols.[18]

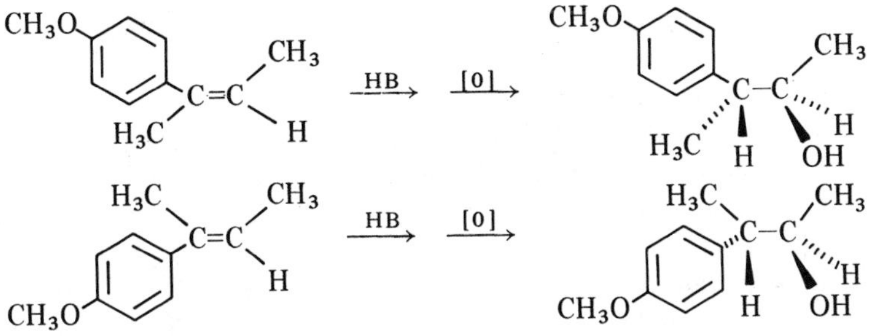

CH_3O CH_3 C=C H_3C H HB [O] CH_3O CH_3 C—C H H_3C H OH

H_3C CH_3 C=C H CH_3O HB [O] H_3C CH_3 C—C H CH_3O H OH

The hydroboration of norbornene proceeds to give *exo*-norborneol almost exclusively[17] **(P2.5)**.

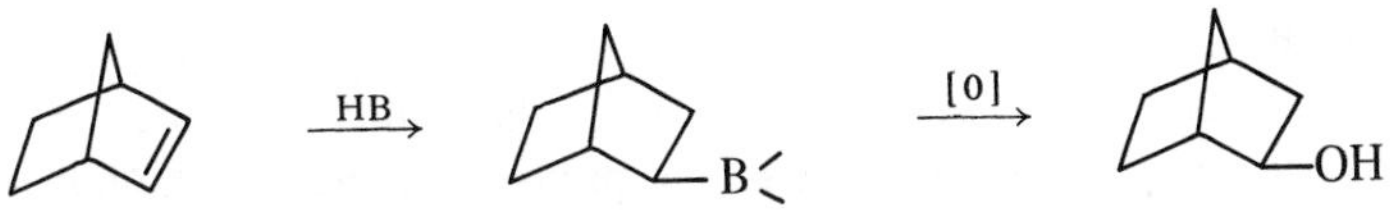

HB B [O] OH

These observations lead to the generalization that hydroboration proceeds by an anti-Markovnikov *cis* addition from the less hindered side of the double bond.[17] It is obviously of major importance to chemists interested in synthesis to be able to predict unambiguously the structure of the hydroboration product. For example, application of the generalization to the hydroboration of α- and β-pinene leads to the prediction that the reaction will take the indicated course.

[17] H. C. Brown and G. Zweifel, *J. Amer. Chem. Soc.*, **81**, 247 (1959).
[18] E. L. Allred, J. Sonnenberg, and S. Winstein, *J. Org. Chem.*, **25**, 26 (1960).

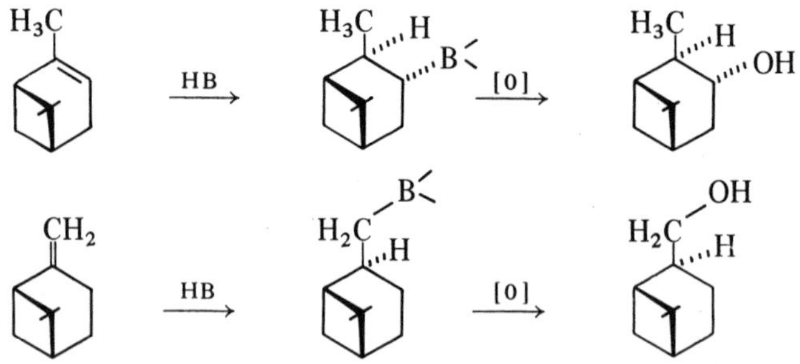

Indeed, the reaction products (**P2.1**, **P2.6**) are fully in accord with the prediction.

1.5. HYDROBORATION OF HINDERED OLEFINS

The reactions of relatively hindered olefins with borane under mild conditions often stop short of the trialkylborane stage. By avoiding the presence of an excess of olefin and by utilizing relatively short reaction times, such reactions can often be controlled to yield the dialkylborane or, in some cases, the monoalkylborane.[19]

Thus the reaction of 2-methyl-2-butene with borane-THF at 0° readily produces bis(3-methyl-2-butyl)borane or disiamylborane (**P2.9**), and the reaction of 2,3-dimethyl-2-butene can be directed quantitatively to the synthesis of the corresponding monoalkylborane, 2,3-dimethyl-2-butylborane or thexylborane (**P2.10**). These derivatives actually exist in tetrahydrofuran solution as the dimers.[20] However, it is convenient to discuss their reactions in terms of the monomeric species in the great majority of cases where the dimeric structure is not an essential part of the reaction.

The hydroboration of cyclopentene proceeds rapidly to the tricyclopentylborane stage, and it is not yet possible to prepare dicyclopentylborane directly. However, the hydroboration of cyclohexene can be readily controlled to yield dicyclohexylborane (**P2.8**). Presumably, both the lower reactivity of cyclohexene and the insolubility of the intermediate, dicyclohexylborane, are factors in its ready synthesis.

Both 1-methylcyclohexene and α-pinene, as trisubstituted olefins, are readily converted into the corresponding dialkylboranes (**P2.1**, **P2.4**).

[19] H. C. Brown and A. W. Moerikofer, *J. Amer. Chem. Soc.*, **84**, 1478 (1962).
[20] H. C. Brown and G. J. Klender, *Inorg. Chem.*, **1**, 204 (1962).

$$2\ (\alpha\text{-pinene}) + BH_3 \longrightarrow (\text{pinanyl})_2BH$$

The reaction with optically active α-pinene makes available an asymmetric hydroborating agent (**P2.1**), whose application will be discussed later (**S3.3**, **P4.9**).

1.6. HYDROBORATION OF DIENES AND ACETYLENES

The hydroboration of dienes with the trifunctional molecule, borane, obviously opens up numerous possibilities for the formation of cyclic and polymeric organoboranes as intermediates.[21] These organoboranes are readily oxidized by alkaline hydrogen peroxide to produce the corresponding glycols.[22]

$$H_2C{=}C(CH_3){-}C(CH_3){=}CH_2 \xrightarrow{HB} \xrightarrow{[O]} HOH_2CCH(CH_3)CH(CH_3)CH_2OH$$

$$H_2C(CH{=}CH_2){-}H_2C(CH{=}CH_2) \xrightarrow{HB} \xrightarrow{[O]} H_2C(CH_2CH_2OH){-}H_2C(CH_2CH_2OH)$$

Consequently, it originally appeared that for synthetic purposes it would be possible to ignore the finer details of the structures of the boron intermediates.[4]

However, certain anomalies were observed which clearly had their origin in these structures. For example, the product from 1,5-hexadiene contains only 69% of the 1,6-diol; 22% of the 1,5- and 9% of the 2,5- are also present. Independent hydroboration of the two terminal double bonds in the usual 1:2 ratio of 94:6 leads to the prediction of 88.36% 1,6-, 11.28% 1,5-, and 0.36% 2,5-. Similarly, 1,5-pentadiene yields only 38% of the 1,5-diol, with 62% of the 1,4-pentanediol present in the product. Clearly the hydroboration must have proceeded via the preferred formation of a five-membered boron heterocycle.[22]

In some cases, the formation of a cyclic organoborane intermediate is readily achieved. For example, the hydroboration of 2,4-dimethyl-1,4-pentadiene and

[21] R. Köster, *Advan. Organometal. Chem.*, **2**, 257 (1964).
[22] G. Zweifel, K. Nagase, and H. C. Brown, *J. Amer. Chem. Soc.*, **84**, 183 (1962).

2,5-dimethyl-1,5-hexadiene can be controlled to yield the corresponding borinane (**P2.12**) and borepane derivatives.[23]

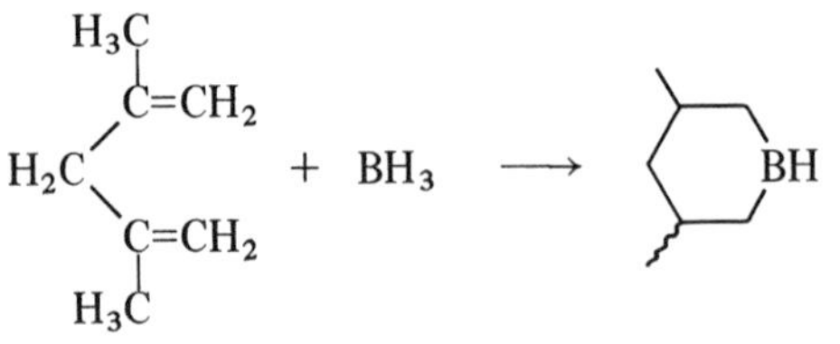

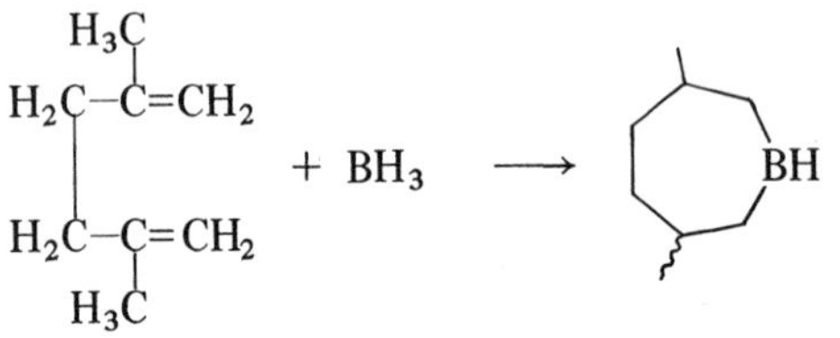

H_3C, $C{=}CH_2$, H_2C, $C{=}CH_2$, H_3C + BH_3 ⟶ BH

H_3C, $H_2C{-}C{=}CH_2$, $H_2C{-}C{=}CH_2$, H_3C + BH_3 ⟶ BH

The hydroboration of 1,5-cyclooctadiene yields an intermediate which undergoes the usual oxidation to yield only *cis*-1,4- and *cis*-1,5-cyclooctanediols.[24] Clearly, bicyclic derivatives, such as 9-borabicyclo[3.3.1]nonane and its [4.2.1] isomer must be involved. Simple thermal treatment (refluxing THF) converts the [4.2.1] isomer into the [3.3.1] derivative, making this interesting dialkylborane readily available[7] (**P2.11**). Oxidation provides *cis*-1,5-cyclooctanediol in high purity (**P2.11**).

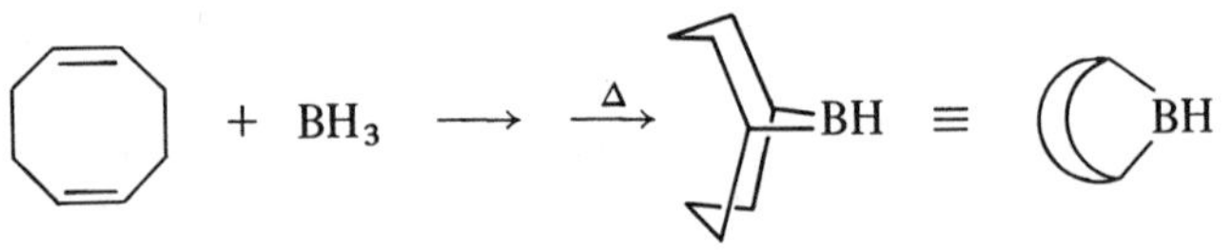

+ BH_3 ⟶ $\xrightarrow{\Delta}$ BH ≡ BH

Utilization of such monofunctional dialkylboranes (**S3**) greatly simplifies the hydroboration of dienes (**P4.2, P4.5**).

Treatment of acetylenes with the theoretical amount of borane results in dihydroboration. The products are predominantly the *gem*-dibora derivatives.[25,26]

[23] E. Negishi and H. C. Brown, *J. Amer. Chem. Soc.*, **95**, 6757 (1973).
[24] E. F. Knights and H. C. Brown, *J. Amer. Chem. Soc.*, **90**, 5280 (1968).
[25] H. C. Brown and G. Zweifel, *J. Amer. Chem. Soc.*, **83**, 3834 (1961).
[26] G. Zweifel and H. Arzoumanian, *J. Amer. Chem. Soc.*, **89**, 291 (1967).

$$CH_3CH_2CH_2C{\equiv}CH \xrightarrow{HB} CH_3CH_2CH_2CH_2CH(B{<})_2$$

The precise structures of these *gem*-dibora derivatives have not yet been worked out.

Monohydroboration of internal acetylenes can be achieved by using controlled amounts of reagent.

$$3\ RC{\equiv}CR + BH_3 \longrightarrow \left(\begin{matrix} R & & R \\ & C{=}C & \\ H & & \end{matrix}\right)_3 B$$

However, under these conditions terminal acetylenes undergo preferential dihydroboration. The use of disubstituted boranes greatly simplifies the problem of achieving the monohydroboration of acetylenes to form the desired vinylboranes (**S3, P4.3, 4.11**).

1.7. HYDROBORATION OF FUNCTIONAL DERIVATIVES

The reactions of diborane with carbon-carbon double and triple bonds are very fast, much faster in fact than the reaction of diborane with many functional groups.[2] An important consequence is that many functional groups can be tolerated in the hydroboration reaction, readily providing organoboranes containing such substituents. For the first time the organic chemist has at his disposal a relatively reactive organometallic with a wide variety of possible substituents.

The presence of relatively inert substituents, such as halogen and alkoxy groupings, generally does not cause any difficulty, and the hydroborations of *p*-chlorostyrene and *p*-methoxystyrene proceed normally.[16] Similarly, the aliphatic derivatives, allyl chloride[27] and vinyl ethyl ether,[28] and a number of related derivatives have been successfully hydroborated.[29,30]

Even for cases in which the molecule contains reducible groups, it has been possible to achieve hydroboration. Thus the methyl ester of 10-undecenoic

[27] M. F. Hawthorne and J. A. Dupont, *J. Amer. Chem. Soc.*, **80**, 5830 (1958).
[28] B. M. Mikhailov and T. A. Shchegoleva, *Izvest. Akad. Nauk SSSR*, 546 (1959).
[29] H. C. Brown and K. A. Keblys, *J. Amer. Chem. Soc.*, **86**, 1791 (1964).
[30] H. C. Brown and O. J. Cope, *J. Amer. Chem. Soc.*, **86**, 1801 (1964).

acid has been converted into the borane and then into the 11-hydroxy derivative via hydroboration-oxidation[31] or into the 11-bromo derivative via brominolysis[32] (**P6.7**). Even derivatives containing the ester grouping much closer to the double bond are readily utilized.[33]

The high reactivity of the carbonyl group in many aldehydes and ketones, as well as that of the free carboxylic acid group,[34] toward diborane would doubtlessly result in a competition for the diborane. However, simple conversion of such groups into the corresponding acetals, ketals, or esters should adequately protect them and permit the desired hydroboration of the carbon-carbon double or triple bonds in the structure.

Electronegative substituents can greatly influence the direction of the addition of the H–B$<$ bond. For example, compare the relative effect of the chlorine and methoxy substituents.[35]

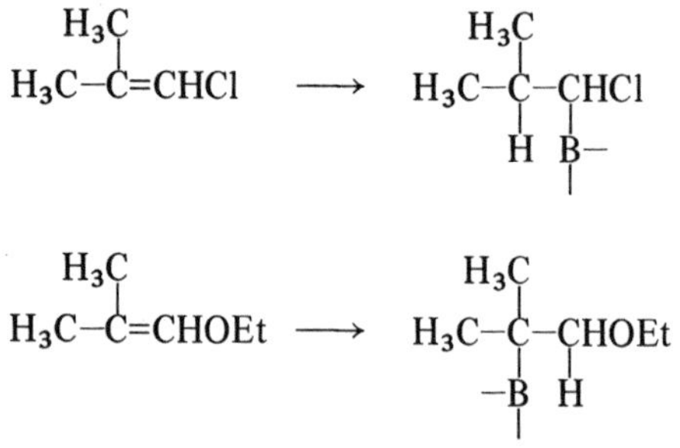

$$H_3C-\overset{H_3C}{\overset{|}{C}}=CHCl \longrightarrow H_3C-\underset{\underset{H}{|}}{\overset{H_3C}{\overset{|}{C}}}-\underset{\underset{-B-}{|}}{C}HCl$$

$$H_3C-\overset{H_3C}{\overset{|}{C}}=CHOEt \longrightarrow H_3C-\underset{\underset{-B-}{|}}{\overset{H_3C}{\overset{|}{C}}}-\underset{\underset{H}{|}}{C}HOEt$$

The opposing electronic effects of the chloro and alkoxy substituents on the observed directive effect in this system parallels that pointed out earlier for the effects of such substituents in the *para* position of styrene.[16]

In some cases, such powerful directive effects may complicate a particular synthesis; in others it may be possible to take advantage of the directive influence to facilitate a particular synthetic route.

A systematic study of the hydroboration of representative 1-butenyl (vinyl) derivatives,[35] 2-butenyl (crotyl) derivatives,[36] 3-butenyl,[37] and related 3-cyclopentenyl derivatives[38] provide the basis for an understanding of the significance of such directive effects.

[31] R. Dulou and Y. Chrétien-Bessière, *Bull. Soc. Chim. France*, 1362 (1959).
[32] H. C. Brown and C. F. Lane, *J. Amer. Chem. Soc.*, **92**, 6660 (1970).
[33] H. C. Brown and K. A. Keblys, *J. Amer. Chem. Soc.*, **86**, 1795 (1964).
[34] H. C. Brown and W. Korytnyk, *J. Amer. Chem. Soc.*, **82**, 3866 (1960).
[35] H. C. Brown and R. L. Sharp, *J. Amer. Chem. Soc.*, **90**, 2915 (1968).
[36] H. C. Brown and R. M. Gallivan, *J. Amer. Chem. Soc.*, **90**, 2906 (1968).
[37] H. C. Brown and M. K. Unni, *J. Amer. Chem. Soc.*, **90**, 2902 (1968).
[38] H. C. Brown and E. F. Knights, *J. Amer. Chem. Soc.*, **90**, 4439 (1968).

For example, as pointed out above, in the 1-butenyl (vinyl) derivatives, the presence of an alkoxy substituent directs the boron strongly to the 2-position. On the other hand, a chlorine substituent directs the boron to the 1-position. Acetoxy lies between. In the 2-butenyl (crotyl) derivatives, the strong directive influence places the boron predominantly (~90%) in the 2-position.

$$CH_3CH{=}CHCH_2X \xrightarrow{HB} CH_3CH_2\underset{\underset{/\backslash}{\mathrm{B}}}{\underset{|}{C}}HCH_2X$$

When X is a good leaving group, such as chloro, elimination occurs rapidly in tetrahydrofuran. The resulting olefin then undergoes rehydroboration. In ethyl ether, however, elimination is slower and the intermediate can be oxidized to the chlorohydrin.[39]

The directive influence of the substituent in the 3-butenyl derivatives can direct as much as 20% of the boron to the 2-position.

$$\underset{\underset{80\%}{\uparrow}}{CH_2}{=}\underset{\underset{20\%}{\uparrow}}{C}HCH_2CH_2Cl$$

Other than this, no complication has been observed. Even this minor difficulty can be circumvented by utilizing the dialkylboranes for the hydroboration (**S3, P4.1**).

The elimination reaction of a β-substituted boron intermediate appears to proceed preferentially *trans* for a halogen substituent and *cis* for an oxygen substituent (alkoxy or acyloxy).[38,39] This makes possible the synthesis of pure isomers in the hydroboration of 3-cyclopentenyl or 3-cyclohexenyl derivatives or related structures. Thus hydroboration-oxidation of isophorone yields only the *trans*-diequatorial-1,2-diol.[40]

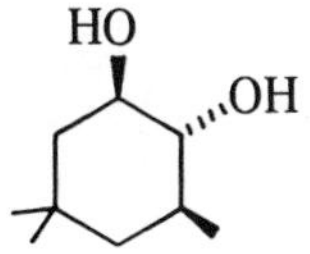

(The minor *cis* hydroboration intermediate undergoes elimination and does not appear in the diol product.)

[39] D. J. Pasto and J. Hickman, *J. Amer. Chem. Soc.*, **90**, 4445 (1968).
[40] J. Klein and E. Dunkelblum, *Tetrahedron Lett.*, 6047 (1966).

In the past, organic chemists have relied heavily on organometallics for the formation of carbon-carbon bonds. However, it was almost axiomatic that a synthesis that relied on an organometallic had to be planned to avoid the presence of a reactive functional substituent either in the organometallic itself or in the substrate undergoing reaction. Now we have the means of achieving a major revolution in this area—the ready formation of carbon-carbon bonds utilizing intermediates with a wide range of functional substituents.

2

HYDROBORATION WITH BORANE: PROCEDURES

2.1. (+)-DIISOPINOCAMPHEYLBORANE [(+)-DI-3-PINANYLBORANE]; (+)-ISOPINOCAMPHEOL [(+)-3-PINANOL]

$$8\ \text{(α-pinene)} + 3\,NaBH_4 + 4\,BF_3{:}OEt_2 \xrightarrow[0^\circ]{DG} 4\ (\text{pinanyl})_2BH + 3\,NaBF_4 + 4\,Et_2O$$

$$(\text{pinanyl})_2BH + H_2O \xrightarrow{25^\circ} (\text{pinanyl})_2BOH + H_2$$

$$(\text{pinanyl})_2BOH + 2\,H_2O_2 + NaOH \xrightarrow{40^\circ} 2\ \text{pinanyl-OH} + NaB(OH)_4$$

Procedure by George Zweifel[1,2]

Procedure

A 300-ml three-necked flask equipped with a condenser, a pressure-equalizing dropping funnel (**N1**), and a mechanical stirrer (**N2**) is assembled (**N3**). The top of the condenser leads to a mercury bubbler. The top of the funnel is fitted with a rubber stopple to permit introduction of materials with the aid of a hypodermic syringe. In the flask are placed 3.1 g (0.080 mole (6% excess)) of sodium borohydride, 100 ml of diglyme (**S9.11**), and 27.2 g (0.200 mole) of (–)-α-pinene (**N4**) (**S9.11**) diluted with 20 ml of diglyme. A hypodermic needle is inserted through the rubber stopple, and the apparatus is flushed with nitrogen. A static nitrogen pressure is maintained through the oxidation stage. The flask is immersed in a water bath (20-25°) and hydroboration is initiated by dropwise addition of 14 ml (0.11 mole (10% excess), d^{25} 1.125) of boron trifluoride etherate (**S9.11**) to the well-stirred reaction mixture over a period of 15 minutes. The (+)-diisopinocampheylborane [(+)-di-3-pinanylborane] precipitates as a white solid as the reaction proceeds. The product can be utilized directly for asymmetric hydroborations (**P4.9**).

The excess hydride is then destroyed by the dropwise addition of 20 ml of water (**N5**), followed by the addition in one portion of 40 ml of aqueous 3*M* sodium hydroxide. The flask is now immersed in an ice-water bath and the borinic acid intermediate is oxidized at 30-50° by the slow addition of 22 ml of aqueous 30% hydrogen peroxide to the well-stirred reaction mixture (**N6, N7**). The mixture is stirred for an additional 30 minutes at room temperature. The reaction mixture is treated with 200 ml of ether, and the ether extract is washed five times with equal volumes of ice water to remove the diglyme. The ether extract is dried over anhydrous magnesium sulfate, and the ether is removed through a short Vigreaux column. The residue is distilled under reduced pressure to separate 26.2 g (85%) of isopinocampheol, bp 80-82° (2 mm). The distillate crystallizes immediately in the collection flask, mp 50-52° (**N8**). Recrystallization from 10 ml of petroleum ether (bp 35-37°) affords 24.3 g (79%) pure (+)-isopinocampheol as needles, mp 55-57°, $[\alpha]^{20}D$ + 32.8° (*c*, 10 in benzene).

Notes

1. The present setup uses simple, unspecialized apparatus (**S9.1, F9.1**).
2. A mechanical stirrer is recommended here because of the formation of the solid intermediate. For many hydroborations on this scale the use of a magnetic stirrer is both adequate and more convenient.
3. The apparatus is dried in an oven and assembled hot under dry nitrogen. Alternatively, the apparatus can be assembled and flame dried in a stream of dry nitrogen (**S9.1**).
4. Commercial (–)-α-pinene, $n^{20}D$ 1.4660, $[\alpha]^{20}D$ -47.9° (**S9.11**).
5. Since the intermediate is present as the dialkylborane, hydrolysis liberates a relatively large quantity of hydrogen (~ 2.5 ℓ). The rate of evolution of hydrogen should be controlled by the slow addition of water and the hydrogen evolved should be conducted to a safe vent. (It is important to hydrolyze the dialkylborane before the hydrogen peroxide is added!)
6. The reaction is exothermic and can be quite vigorous. It should be controlled

by the slow, dropwise addition of hydrogen peroxide. The oxidation utilizes 1 mole of sodium hydroxide per mole of organoborane and 1 mole of hydrogen peroxide (30% hydrogen peroxide $\approx 10M$) per mole of carbon-boron bonds (**S5.7**). (In the present procedure a somewhat larger quantity of sodium hydroxide is used to compensate for residual boron trifluoride that may be present.)

7. The nitrogen atmosphere must be retained until after the oxidation. Oxidation of organoboranes by elemental oxygen, either from the atmosphere or from thermal decomposition of hydrogen peroxide is not stereospecific (**S5.9**).
8. For the distillation of the isopinocampheol the Vigreaux column is attached to an air condenser. The receiver is immersed in an ice bath.

[1] H. C. Brown, N. R. Ayyangar, and G. Zweifel, *J. Amer. Chem. Soc.*, **86**, 393 (1964).
[2] G. Zweifel and H. C. Brown, *Org. Syn.*, **52**, 59 (1972).

2.2. 2-PHENYL-1-PROPANOL

$$12\ C_6H_5C(CH_3)=CH_2 + 3\ NaBH_4 + 4\ BF_3{:}OEt_2 \xrightarrow[25^\circ]{THF} 4\ (C_6H_5CH(CH_3)CH_2-)_3B + 3\ NaBF_4 + 4\ Et_2O$$

$$(C_6H_5CH(CH_3)CH_2-)_3B + 3\ H_2O_2 + NaOH \xrightarrow[30\text{-}40^\circ]{THF} 3\ C_6H_5CH(CH_3)CH_2OH + NaB(OH)_4$$

Procedure by George Zweifel and John R. Schwier[1]

Procedure

A 500-ml flask and accessories are assembled as described in **P2.1**. In the flask are placed 3.1 g (0.080 mole) of sodium borohydride. A hypodermic needle is inserted into a rubber septum at the top of the dropping funnel and a stream of dry nitrogen introduced to flush the apparatus. A static nitrogen pressure is then maintained through the oxidation stage. The reagents are now introduced into the dropping funnel and then to the flask with the aid of suitable syringes (**S9.2**). α-Methylstyrene (**N1**), 35.4 g (0.300 mole), is introduced, followed by

100 ml of dry tetrahydrofuran (**S9.11**). The flask is immersed in a water bath 20-25°), and hydroboration is initiated by the dropwise addition of 14.0 ml (0.11 mole) of boron trifluoride etherate (**S9.11**) to the well-stirred suspension over a period of 1 hour, while the temperature is maintained at 25°. The reaction mixture is maintained for an additional hour at 25°. Then 20 ml of water is added to destroy residual hydride (**N2**). The organoborane is oxidized at 30-40° (water bath) by the addition of 33 ml of a 3*M* solution of sodium hydroxide, followed by the careful dropwise addition of 33 ml of 30% hydrogen peroxide (**P2.1:N6**). Sodium chloride is added to saturate the aqueous phase. The upper tetrahydrofuran layer is separated, washed with saturated, aqueous sodium chloride, and dried over anhydrous magnesium sulfate. Distillation provides 38.8 g (95%) of 2-phenyl-1-propanol, bp 110-112° (13 mm), $n^{20}\text{D}$ 1.5264.

Notes

1. Freshly distilled, bp 60° at 17 mm, $n^{20}\text{D}$ 1.5385, (**S9.11**).
2. The amount of hydrogen evolved will be much smaller than in **P2.1**, since the present reaction proceeds to the R_3B stage.

[1] Adapted from the procedure reported for 4-methyl-1-pentanol: G. Zweifel and H. C. Brown, *Org. React.*, **13**, 1 (1963).

2.3. BORANE-TETRAHYDROFURAN

$$3\,NaBH_4 + 4(C_2H_5)_2O{:}BF_3 \xrightarrow[25^\circ]{DG} 2\,B_2H_6\uparrow + 3\,NaBF_4 + 4(C_2H_5)_2O$$

$$B_2H_6 + 2\,C_4H_8O \xrightarrow{0^\circ} 2\,C_4H_8O{:}BH_3$$

Procedure by M. Mark Midland[1]

Procedure

The apparatus (**F2.1**) is assembled in a hood as follows. A 2-ℓ three-necked flask is fitted with a 500-ml pressure-equalizing funnel, mechanical stirrer, and Dry-Ice condenser (**N1**). The top of the Dry-Ice condenser is connected to two traps in series (**N2**). A rubber serum stopple is placed over the exit of the second trap. A mercury bubbler is connected to the exit by a Tygon tube and a syringe needle inserted into the stopple. A safety valve, consisting of a T-tube immersed in a mercury pool containing acetone over the mercury (to destroy escaping diborane) is placed on the three-way stopcock (**N3**). The system is flame dried

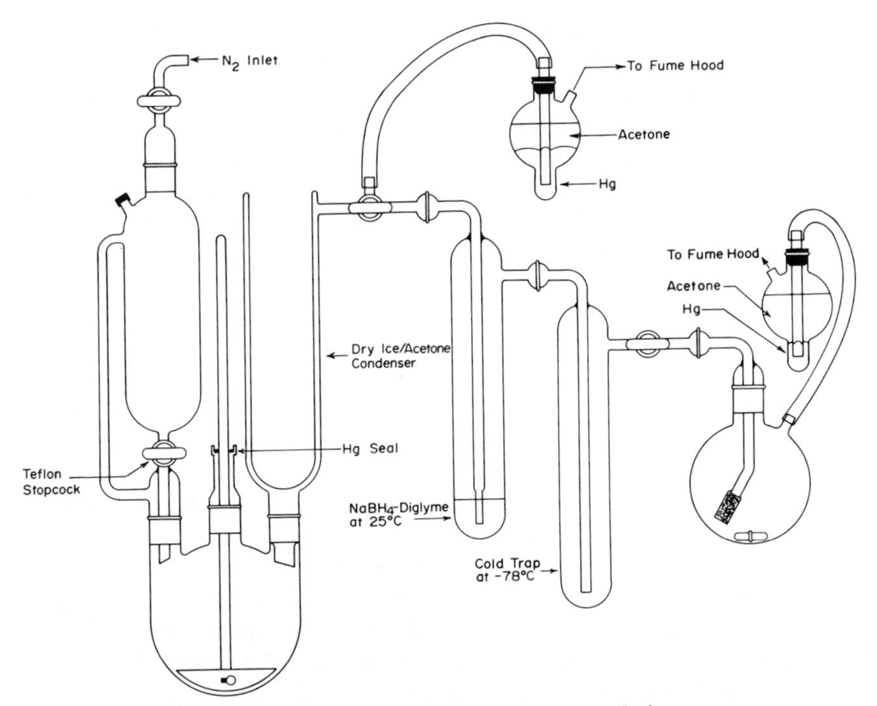

Figure 2.1. All-glass apparatus for the preparation of borane solutions.

while being flushed with nitrogen. After drying, the rubber stopple is replaced by a coarse sintered-glass dispersion tube **(N4)**. A 2-ℓ flask fitted with a sidearm is flushed with nitrogen and charged with 1 ℓ of tetrahydrofuran **(S9.11)**. The sintered-glass dispersion tube is inserted into the 2-ℓ flask and the sidearm connected to a second mercury bubbler containing acetone. The three-necked flask is charged under a blanket of nitrogen with 74.1 g (1.9 mole) of sodium borohydride by replacing the addition funnel temporarily with a powder funnel. The nitrogen flushing is continued. Approximately 500 ml of dry diglyme **(S9.11)** is added to the flask either by a 100-ml syringe or by a double-ended needle **(S9.2) (N5)**. The 2-ℓ flask is cooled by immersion in an ice-salt bath while maintaining the flow of nitrogen. Into the first trap is placed 1 g of sodium borohydride (to remove traces of boron trifluoride or other volatile acids) and sufficient dry diglyme to cover 2 to 3 cm of the central tube. The second trap is cooled in a Dry-Ice acetone bath. The pressure-equalizing funnel is filled with 315 ml (2.5 mole) of boron trifluoride etherate **(S9.11)** with the aid of a 100-ml syringe or double-ended needle **(N5)**. Diborane is generated by the dropwise addition of the boron trifluoride etherate over a 2-3 hour period. A gentle flow of nitrogen is maintained until the generation of diborane has begun (after approximately half of the boron trifluoride has been added) **(N6)**. After the addition is complete, the reaction flask is heated to 60° with a water bath for 1 hour to drive the remaining diborane from the diglyme into the tetrahydrofuran solution **(N7)**. The system is then flushed with nitrogen for 0.5 hr. The sintered-glass tube is removed from the tetrahydrofuran under a stream of nitrogen and replaced with a glass stopper. The sidearm exit to the bubbler is capped with a rubber septum **(N8)**. The yield of diborane is approximately 90%. The resulting solution is approximately 2*M* in borane (BH_3). It can be standardized by measuring the hydrogen evolved on hydrolysis **(S9.10)** **(N9)**.

Properties

Earlier reports have stressed the hazardous nature of diborane. However, during the many years over which the hydroboration reaction has been explored in our laboratories, no difficulties have been encountered in the preparation and handling of such solutions of borane-tetrahydrofuran. Even when deliberately exposed to air, such solutions did not inflame spontaneously, but were slowly oxidized. The diborane solution reacts rapidly with water and should therefore also be protected from exposure to atmospheric moisture. It is thus desirable that solutions be stored under nitrogen and transferred with minimum exposure to oxygen and water, by means of syringes or double-ended needles. The borane solution is stable for several months when kept at 0° under nitrogen. Finally, we have never encountered any health problems in working with such borane solutions. However, it is desirable to maintain the usual precautions in working with a reactive substance whose physiological characteristics have not yet been defined.

Notes

1. All joints and syringes are lubricated with a petroleum jelly, such as Amojell **(S9.2)**.
2. Either short sections of Tygon tubing or ball joints may be used for the connections.
3. Sufficient mercury must be placed in the trap to withstand the pressure drop across the sintered-glass tube.
4. The sintered-glass tube and 2-ℓ flask must be oven-dried at 125° for 12 hours prior to use. It is very important that this sintered tube be completely free of water. Otherwise the hydrolysis of diborane will produce boric acid which could plug the pores and prevent the gas from reaching the solvent.
5. Double-ended needles are available from Aldrich Chemical Company.
6. The initial reaction involves formation of a $NaBH_4 \cdot BH_3$ addition compound.

$$7\ NaBH_4 + 4\ BF_3{:}OEt_2 \longrightarrow 4\ NaBH_4{\cdot}BH_3 + 3\ NaBF_4 + 4\ Et_2O$$

Diborane is not liberated until the addition of boron trifluoride etherate proceeds beyond this point.

7. The solubility of sodium fluoborate decreases at the higher temperatures and the salt separates. As a result, the mechanical stirring must be continuous to avoid "freezing" of the solution. It is desirable that the mechanical stirrer be of the heavy-duty type.
8. For storage over longer periods, a stopcock (fitted with a Teflon plug) sealed to the sidearm is preferable (**S9.9**).
9. Standardized solutions of the borane-tetrahydrofuran complex (1.0*M*) are now available from the Aldrich Chemical Company. These contain a small quantity of sodium borohydride as a stabilizer.

[1]Adapted from the procedure described by H. C. Brown and R. L. Sharp, *J. Amer. Chem. Soc.*, 90, 2915 (1968). The all-glass apparatus was designed by E. F. Knights, S. P. Rhodes, and P. L. Burke.

2.4. *TRANS*-2-METHYLCYCLOHEXANOL

$$2\ \text{1-methylcyclohexene} + H_3B{:}THF \xrightarrow[0^\circ]{THF} (\text{trans-2-methylcyclohexyl})_2BH$$

$$(\text{trans-2-methylcyclohexyl})_2BH + H_2O \xrightarrow{25^\circ} (\text{trans-2-methylcyclohexyl})_2BOH + H_2$$

$$(\text{trans-2-methylcyclohexyl})_2BOH + 2\ H_2O_2 + NaOH \xrightarrow{40^\circ} 2\ \text{trans-2-methylcyclohexanol} + NaB(OH)_4$$

Procedure by George Zweifel[1] and Akira Suzuki[2]

Procedure

A dry 500-ml 3-neck flask fitted with a septum inlet, condenser, pressure equalizing dropping funnel, magnetic stirring bar, and mercury bubbler is assembled, dried, flushed with nitrogen, and maintained under a static pressure until following the oxidation (**N1**). In the flask are placed 28.8 g (0.300 mole) of 1-methylcyclohexene **(N2)** and 150 ml of dry tetrahydrofuran **(S9.11)**. The flask is immersed in an ice-bath, and 75ml of 2.00M borane-tetrahydrofuran complex **(P2.3)(N3)** is added dropwise over a period of 0.5 hour to the stirred solution. The dialkylborane forms and precipitates as a white solid. The reaction mixture is permitted to stir for an additional hour at room temperature to ensure completion of the reaction. Water, 15 ml, is added slowly over several minutes, and the mixture is allowed to stir at room temperature until hydrogen is no longer evolved **(P2.1:N5)**. The flask is immersed in an ice-water bath, and 50 ml of 3*M* sodium hydroxide is rapidly added to the reaction mixture. The organoborinic acid intermediate is now oxidized by the slow careful dropwise addition of 33 ml of 30% hydrogen peroxide **(P2.1:N6, P2.1:N7)** at a rate such that the temperature is maintained at approximately 40°. The reaction mixture is now allowed to stir for an additional hour at 50° to ensure completion of the oxidation. The mixture is brought to room temperature, and sodium chloride is added to saturate the lower aqueous phase **(N4)**. The tetrahydrofuran phase is separated and dried over anhydrous magnesium sulfate. Distillation provides 29.1 g of *trans*-2-methylcyclohexanol, bp 166° at 741 mm, n^{20}D 1.4611, a yield of 85% **(N5)**.

Notes

1. The present setup uses the more specialized equipment described in **S9.1**, **F9.4**. However, the preparation can be carried out with simple, unspecialized pieces, as described in **P2.1** **(S9.1, F9.1)**.
2. 1-Methylcyclohexene, bp 108° at 743 mm, n^{20}D 1.4508 **(S9.11)**, is made by dehydration with iodine of 1-methylcyclohexanol, prepared from cyclohexanone and methyl magnesium iodide. It is now commercially available.
3. Alternatively, the 1.0*M* borane-tetrahydrofuran complex available from the Aldrich Chemical Company should be applicable. This solution contains a small amount of sodium borohydride as a stabilizing agent. It has been shown to be applicable in a representative number of preparations, but has not been tested for all of the procedures here reported. These have generally utilized borane prepared as described in **P2.3**.
4. In cases where the product is highly water soluble, such as a glycol, potassium carbonate is used to saturate the aqueous phase **(P2.11, P6.3)**.
5. 1-Methylcyclopentene,[1] 1-phenylcyclopentene,[1] 1-phenylcyclohexene,[1] α-pinene **(P2.1)**, Δ^3-carene,[2] α-thujene,[3] α-cedrene,[4] and thujopsene[5] have all been hydroborated and converted to the related, isomerically pure *trans* alcohols.

[1] H. C. Brown and G. Zweifel, *J. Amer. Chem. Soc.*, 83, 2544 (1961).
[2] H. C. Brown and A. Suzuki, *J. Amer. Chem. Soc.*, 89, 1933 (1967).

[3]S. P. Acharya, H. C. Brown, A. Suzuki, S. Nozawa, and M. Itoh, *J. Org. Chem.*, **34**, 3015 (1969).
[4]S. P. Acharya and H. C. Brown, *J. Org. Chem.*, **35**, 196 (1970).
[5]S. P. Acharya and H. C. Brown, *J. Org. Chem.*, **35**, 3874 (1970).

2.5. *EXO*-NORBORNEOL

3 (norbornene) + H_3B:THF $\xrightarrow[10^\circ]{THF}$ (norbornyl)$_3$B

(norbornyl)$_3$B + 3 H_2O_2 + NaOH $\xrightarrow{40^\circ}$ 3 (norbornyl)OH + $NaB(OH)_4$

Procedure by Clinton F. Lane[1]

Procedure

A 250-ml three-necked, round-bottomed flask equipped with a reflux condenser, thermometer, magnetic stirring bar, and pressure-equalizing dropping funnel is assembled (**S9.1**, **F9.1**). The apparatus is flamed with a Bunsen burner while flushing the system with dry nitrogen (**N1**). The nitrogen stream is introduced through a septum inlet (**N2**) on the dropping funnel using a hypodermic needle which has been wired into a nitrogen line. The nitrogen is vented through a mercury or mineral oil bubbler connected to the outlet of the condenser. After cooling to room temperature under a positive nitrogen pressure, the reaction flask is charged with 18.8 g (0.200 mole) of norbornene by removing the condenser and adding the solid norbornene as quickly as possible under a blanket of nitrogen (**N3**). The system is then reflushed with nitrogen for 1 minute. Tetrahydrofuran (**N4**), 10 ml, is added via a dry syringe, and the clear, colorless solution is cooled to 5-10° with stirring in an ice-water bath. The borane-tetrahydrofuran (BH_3:THF) complex (73 ml of a 1.0*M* THF solution, 0.067 mole plus 10% excess) is added to the calibrated dropping funnel via a double-ended needle (**N2, N5**).

Hydroboration is achieved by the slow, dropwise addition of the BH_3:THF solution to the norbornene-THF solution (**N6**). Following the addition, the clear, colorless reaction mixture is stirred at room temperature for 0.5 hour to complete the reaction. Excess hydride is destroyed by the careful addition of 5 ml of water (**N7**). After hydrogen is no longer evolved, approximately 10 minutes, 25 ml of 3*M* aqueous sodium hydroxide is added to the reaction mixture (**P2.1:N7**). Hydrogen peroxide, 25 ml of a 30% aqueous solution, is introduced into the dropping funnel and added dropwise to the stirred reaction mixture at a rate such that the temperature of the reaction mixture does not exceed approximately 40° (**P2.1:N6**). When the addition is complete, the reaction mixture is heated at 50° and maintained there for 1 hour to ensure complete oxidation.

Isolation of the *exo*-norborneol (100% yield by VPC) is accomplished by pouring the cooled, two-phase reaction mixture into a separatory funnel, adding 50 ml of ethyl ether, and removing the lower aqueous layer. The organic layer is washed with water (2 × 25 ml) and saturated aqueous sodium chloride (1 × 25 ml). The organic layer is dried over anhydrous potassium carbonate, filtered, and concentrated on a rotary evaporator to give 19-21 g (85-93% based on norbornene) of a white solid, mp 116-119° (**N8**). The norborneol isolated in this manner has a purity of at least 98% by VPC analysis (1/8 in. × 8 ft column, 5% Carbowax 20M on Varaport 30). The isomeric purity is >99% *exo*- by either VPC or NMR analysis. If desired, *exo*-norborneol of >99% purity (mp 126°, lit.[2] 126-127°) may be obtained by sublimation.

Notes

1. Alternatively, the apparatus may be dried in an oven at 125°, assembled hot, and then flushed with nitrogen (**S9.1**).
2. Rubber serum stoppers, double-ended stainless steel needles, and the borane-tetrahydrofuran complex are available from the Aldrich Chemical Company.
3. The bulletin, *Handling Air-Sensitive Solutions,* available upon request from the Aldrich Chemical Company, Milwaukee, Wisconsin 53233, is helpful. See also **S9**.
4. Tetrahydrofuran is dried by storing under nitrogen over calcium hydride or 4 A molecular sieves. See also **S9.11**.
5. Alternatively, the dry THF and the BH_3:THF solutions may be conveniently added to the reaction vessel through a rubber septum inlet using syringes (**S9.1, S9.2**).
6. The hydroboration reaction is exothermic, proceeding readily to the trialkylborane stage, in contrast to **P2.4**.
7. The addition of water should be controlled so as to avoid a rapid evolution of hydrogen. The hydrogen evolved, approximately 0.5 ℓ, should be safely vented.
8. Care must be taken to avoid loss of product due to sublimation when using the rotary evaporator.

[1] C. F. Lane, *Aldrichimica Acta,* **6**, 36 (1973).
[2] H. C. Brown and G. Zweifel, *J. Amer. Chem. Soc.,* **83**, 2544 (1961).

2.6. (−)-*CIS*-MYRTANOL

$$3\ (\text{CH}_2{=}) \ + \ H_3B{:}S(CH_3)_2 \xrightarrow[0^\circ]{\text{Hexane}} (\text{CH}_2-)_3B \ + \ (CH_3)_2S$$

$$(\text{CH}_2-)_3B \ + \ 3\,H_2O_2 \ + \ NaOH \xrightarrow[40^\circ]{\text{EtOH}} 3\ \text{CH}_2OH \ + \ NaB(OH)_4$$

Procedure by Clinton F. Lane[1]

Procedure

A dry 2-ℓ flask equipped with a mechanical stirrer, pressure-equalizing dropping funnel, and reflux condenser is flushed with dry nitrogen and maintained under positive nitrogen pressure (**S9.1**, **F9.12**). The flask is then charged with 204 g (238 ml 1.5 mole) of (−)-β-pinene (**N1**) and 500 ml of hexane and cooled to 0-5° with an ice-water bath. Hydroboration is achieved by the dropwise addition of 52.5 ml (0.55 mole) of borane-methyl sulfide complex (BMS) (**N1**) (**P2.8:N1**). Following addition of the hydride (0.5 hour), the cooling bath is removed and the solution stirred for 3 hours at 25°. Ethanol, 500 ml (*CAUTION:* gas evolution), is then added (**N2**) followed by 185 ml of 3*M* aqueous sodium hydroxide. The reaction mixture is immersed in an ice-water bath and hydrogen peroxide (185 ml of 30% aqueous solution) is added at such a rate that the temperature does not go above approximately 40° (**P2.1:N6**). Immediately following the addition of the peroxide (~1 hour), the cooling bath is removed, and the reaction mixture is heated at 50° for 1 hour. The reaction mixture is then poured into 5 ℓ of ice water. After adding 2 ℓ of ether and mixing thoroughly, the lower aqueous layer is removed and discarded. The upper organic layer is washed with water (2 × 1 ℓ), washed with saturated aqueous sodium chloride, 1 ℓ, dried over anhydrous potassium carbonate, filtered, and concentrated on a rotary evaporator to give 230 g of a light yellow oil. Short-path vacuum distillation of this oil gives 196 g (85%) of (−)-*cis*-myrtanol, purity >98% by VPC analysis, bp 65-67° (0.2 mm), n^{20}_D 1.4911, $[\alpha]^{22}_D$ -19.5° [lit.[2] bp 70-72° (1 mm), n^{20}_D 1.4910, $[\alpha]^{25}_D$ -21°].

Notes

1. All starting materials, including BMS, were used directly as obtained from

the Aldrich Chemical Company.

2. The ethanol is required as a cosolvent to facilitate the oxidation. It is occasionally required even in tetrahydrofuran systems for the oxidation of relatively resistant boron intermediates (**S5.7, P6.10**).

[1]C. F. Lane, *J. Org. Chem.*, **39**, 1437 (1974).
[2]G. Zweifel and H. C. Brown, *J. Amer. Chem. Soc.*, **86**, 393 (1964).

2.7. TRI-*SEC*-BUTYLBORANE

$$3\ CH_3CH{=}CHCH_3 + H_3B{:}THF \xrightarrow[0^\circ]{THF} (CH_3CH_2\overset{\overset{\displaystyle CH_3}{|}}{C}H-)_3B$$

Procedure by M. Mark Midland and S. Krishnamurthy

Procedure

A dry 250-ml flask equipped with a septum inlet, reflux condenser, magnetic stirring bar, and a mercury bubbler (attached to the top of the condenser) is flushed with nitrogen introduced through the septum inlet. The needle is removed and the system maintained under a static pressure of nitrogen (**S9.1**). The flask is charged by means of a 100-ml syringe with 0.100 mole (50 ml of a 2.00*M* solution) of borane in tetrahydrofuran (**P2.3**)(**P2.4:N3**), and the flask is cooled to 0° with an ice-water bath (**N1**). A cylinder of 2-butene is placed on a balance adjacent to the reaction flask (**S9.3**). A Tygon tube fitted to a syringe needle is attached to the cylinder of 2-butene; the tube and needle flushed with 2-butene. The weight of the cylinder and its contents is noted. The needle is then inserted through the septum inlet of the reaction flask to a point just above the unstirred solution. The valve of the cylinder is gently opened and a slow stream of gas is passed into the flask. The tip of the needle is now submerged into the reaction mixture, and stirring is initiated (**N2**). The amount of 2-butene introduced is monitored by the weight loss of the cylinder. A slight excess over the 16.8 g (0.300 mole) required is added to compensate for the uncertainties in the weighing procedure. After the addition has been completed, the needle is removed and the cylinder valve closed. The reaction mixture is stirred for 10 minutes to ensure complete hydroboration. Any excess 2-butene may be removed, if desired, by warming the solution to 50° while passing a slow stream of nitrogen over the reaction mixture. The yield of tri-*sec*-butylborane is essentially quantitative.

For many purposes the tetrahydrofuran solution may be used directly, as for the preparation of lithium[1] or potassium tri-*sec*-butylborohydride[2] (**N3**). If tri-*sec*-butylborane is desired free of tetrahydrofuran, the flask is connected through a stopcock to a water aspirator. The flask is warmed in a 40° water bath until the last of the tetrahydrofuran has been removed. The stopcock is then closed and nitrogen introduced to release the vacuum. This material is quite pure and entirely satisfactory for most applications (**N4**).

To obtain a distilled product, the following procedure is utilized. The flask is connected in a stream of nitrogen to a Claisen head, connected to a rotating fraction collector (**S9.7**). The flask used to collect the main fraction is fitted with a sidearm containing a Teflon stopcock and a septum inlet (**N5**). Distillation is carried out at 3 mm using a slow nitrogen bleed (**N6**). A 2-3 ml forerun is collected, followed by the main fraction, bp 63-64° at 3 mm. After the distillation is complete, the vacuum is released with nitrogen. A slow stream of nitrogen is passed into the flask through the septum. The flask is quickly removed from the distillation apparatus and closed by a glass stopper. The yield of product is 80-90% (**N7, N8**).

Notes

1. **P2.3:N8**. Alternatively, the hydroboration of 2-butene in ethyl ether with borane-methyl sulfide complex (**P2.6**) would offer some advantages in isolating the product, since these solvents are highly volatile (bp 35°) (**P2.8**).
2. A slow stream of nitrogen is passed through the mercury bubbler to prevent backing up of the mercury when the stirrer is started.
3. Both lithium tri-*sec*-butylborohydride[1] and potassium tri-*sec*-butylborohydride[2] possess exceptional characteristics for stereoselective reductions. They are now commercially available from the Aldrich Chemical Company as the reagents L-Selectride and K-Selectride.
4. The organoboranes react vigorously with oxygen, the lower alkylboranes being spontaneously flammable in air. Tri-*sec*-butylborane does not inflame, but smokes when exposed to air. Dialkylboranes, with the exception of solid 9-BBN (**P2.11**), are even more reactive. Consequently, organoboranes should be maintained under an inert atmosphere. They may be stored indefinitely at room temperature in flasks fitted with Teflon stopcocks which are capped by rubber septa. To remove a sample, the Teflon stopcock is opened, and a hypodermic syringe is used to penetrate the rubber serum cap and the bore of the open stopcock. When the syringe is removed, the stopcock is again closed (**S9.2, S9.9**).
5. This receiver will then be used for storage (see **N4**). Alternatively, it can be collected in a flask with an ordinary septum inlet and then transferred to a storage flask (**S9.9**).

6. The tri-*sec*-butylborane must not be overheated, since the 2-butyl moieties can undergo isomerization to 1-butyl moieties at elevated temperatures (**S5.1**). Consequently, the pot temperature should not exceed 100°.
7. The purity of the tri-*sec*-butylborane may be determined by VPC examination (**S9.10**) or by oxidation (**S9.10**). For the latter, a syringe is used to place a weighed 1.0-ml sample into a nitrogen flushed 50-ml flask equipped with a septum inlet, a magnetic stirring bar, and a reflux condenser. The flask is charged by syringe with 10 ml of tetrahydrofuran and 1.5 ml of 3*M* sodium hydroxide. The flask is immersed in a water bath, and 1.5 ml of 30% hydrogen peroxide is added dropwise at such a rate as to maintain the temperature at 40-50°. After the addition is complete, the reaction is heated at 50° for 10 minutes. The aqueous phase is saturated with potassium carbonate, and a weighed 1-ml sample of *n*-dodecane is added as an internal standard. The organic layer is analyzed in a ¼ in. × 6 ft 10% Carbowax 20M column. A 98% yield of 2-butanol with no 1-butanol was realized in the analysis of the above preparation.
8. The above procedure is applicable to the synthesis of a wide variety of organoboranes. However, the higher trialkylboranes may undergo both isomerization and thermal decomposition upon distillation (**S5.1**, **S5.4**). Consequently, these must be purified by other methods, such as crystallization, or used directly.

[1] H. C. Brown and S. Krishnamurthy, *J. Amer. Chem. Soc.*, **94**, 7159 (1972).
[2] C. A. Brown, *J. Amer. Chem. Soc.*, **95**, 4100 (1973).

2.8. DICYCLOHEXYLBORANE

$$2\ \text{cyclohexene} + H_3B{:}S(CH_3)_2 \xrightarrow{EE} (\text{cyclohexyl})_2BH + (CH_3)_2S$$

Procedure by Gerald J. Klender[1] and Surendra U. Kulkarni

Procedure

A dry two-neck 250-ml flask (weighed with stoppers) is equipped with a magnetic stirring bar, pressure-equalized dropping funnel fitted with a rubber septum, and a reflux condenser leading to a mercury bubbler. The apparatus is flushed with dry nitrogen introduced through a hypodermic needle penetrating the rubber septum, and the apparatus is then maintained under a positive pressure of nitrogen (**S9.1**, **F9.1**). The flask is charged with 16.4 g (0.200 mole) of cyclohexene (**S9.11**), followed by 75 ml of anhydrous ethyl ether (**S9.11**). The flask is immersed in an ice-water bath, and 7.7 g (0.100 mole) of borane-methyl sulfide complex (**N1**) is added over a period of 0.5 hour, followed by an additional 25 ml of ethyl ether (to achieve a quantitative transfer of the reagent). The solution is stirred for 3 hours at 0°. Dicyclohexylborane precipitates as a white solid (**N2, N3**). The dicyclohexylborane can be readily freed from the volatile solvents, ethyl ether (bp 35°) and methyl sulfide (bp 35°), by distilling them in a slow stream of nitrogen. The yield is quantitative. Crystalline dicyclohexylborane has been purified by sublimination under vacuum, mp 103-105°.[1]

Notes

1. The borane-methyl sulfide complex, d^{25} 0.80, is conveniently introduced with the aid of a hypodermic syringe: 9.6 ml = 7.7 g.
2. Dicyclohexylborane can be used *in situ* to hydroborate other olefins.[2] Thus treatment with 1-octene yields dicyclohexyl-*n*-octylborane.[3] Similarly, it has been used to hydroborate various unsaturated derivatives containing functional groups, such as ethyl vinylacetate and allyl cyanide[3] and to monohydroborate acetylenes (**P8.17**). For such applications it can be conveniently prepared with borane-tetrahydrofuran following **P2.9** and utilized directly without isolation.
3. The dicyclohexylborane suspended in ethyl ether exhibits no significant change in 6 hours at 25° and less than 5 percent in 24 hours. It is slightly less stable in tetrahydrofuran. For best results the solution should be used immediately after its preparation. The solid dicyclohexylborane is considerably more stable. It exhibits no significant change in 15 days at 0°.

[1] H. C. Brown and G. J. Klender, *Inorg. Chem.*, **1**, 204 (1962). The original procedure utilized diborane in diethyl ether.
[2] G. Zweifel, N. R. Ayyangar, and H. C. Brown, *J. Amer. Chem. Soc.*, **85**, 2072 (1963).
[3] H. C. Brown, G. W. Kabalka, and M. W. Rathke, *J. Amer. Chem. Soc.*, **89**, 4530 (1967).

2.9. BIS-(3-METHYL-2-BUTYL)BORANE [DISIAMYLBORANE]

$$2\ (H_3C)_2C{=}CH(CH_3) + BH_3{:}THF \xrightarrow{THF} (H{-}C(CH_3)_2{-}CH(CH_3){-})_2BH + THF$$

Procedure by Andrew W. Moerikofer[1]

Procedure

A carefully dried 200-ml reaction flask, fitted with a septum inlet and a magnetic stirring bar, is attached to a mercury bubbler (**S9.1**, **F9.5**). The flask is flushed with dry nitrogen and then maintained under a static pressure of nitrogen (**N1**). In the flask is placed 50.0 ml of a 2.00*M* solution of borane-tetrahydrofuran (**P2.3**). The flask is immersed in an ice-salt bath, and the temperature is lowered to -10 to -15°(**N2**). A hypodermic syringe is used to measure out 50 ml of a 4.00*M* solution of 2-methyl-2-butene (**N3**), and the solution is then added slowly through the septum inlet to the stirred borane reagent over a period of approximately 30 minutes, maintaining the temperature at approximately 0° (**N4**). The reaction mixture is allowed to stand for 2 hours to complete the reaction. A clear solution of disiamylborane, 1.00*M*, in tetrahydrofuran is thus obtained (**N5**). It can be utilized *in situ* for hydroborations (**S3.6**) or reductions. Alternatively, aliquots can be removed with a hypodermic syringe and utilized (**N6, N7**). Such aliquots can be analyzed for active hydrogen by hydrolysis (**S9.10**) and for the presence of two unisomerized siamyl groups (as 3-methyl-2-butanol) after oxidation (**S9.10, P2.7:N7**) (**N8**).

Notes

1. The present apparatus is simplified over those described earlier (**P2.1,P2.5**)—it does not utilize a dropping funnel or reflux condenser.
2. A major difficulty in achieving the quantitative synthesis of disiamylborane is the high volatility of 2-methyl-2-butene. The low temperature used (-10 to -15°) minimized losses of the olefin.
3. Commercial 99%: n^{20}D 1.3890 (**S9.11**).
4. It proved convenient to prepare a large quantity of this standard solution of 2-methyl-2-butene in tetrahydrofuran. This solution, stored at 0° in a cold room along with the standard borane-tetrahydrofuran solution, permits rapid, convenient preparation of disiamylborane as required.
5. Disiamylborane actually exists as a dimer in tetrahydrofuran.[2] The product has been isolated and described.[2] However, it is more conveniently handled in the form of the solution. The kinetics of the reaction of disiamylborane (dimer) with representative olefins has been examined.[1]
6. Such solutions should not be stored for more than a few hours at 0°. It has proved quite convenient to prepare the reagent fresh as required by merely mixing aliquots of the two standard solutions as described above.
7. In the original procedure, the borane solution was added to the 3-methyl-2-butene at low temperature. The reaction of disiamylborane with a third molecule of the olefin is sufficiently slow at the reaction temperature so that no difficulty is normally encountered with the procedure. However, the present order of mixing the reagents minimizes such possible formation of trisiamylborane in the initial stages of the reaction.

8. The 1.0*M* commercial solution of the borane-tetrahydrofuran complex (Aldrich) can be used. For most purposes, the more dilute solution of disiamylborane thus obtained is satisfactory. It can be concentrated by pumping off excess tetrahydrofuran at room temperature.

[1] H. C. Brown and A. W. Moerikofer, *J. Amer. Chem. Soc.*, **83**, 3417 (1961).
[2] H. C. Brown and G. J. Klender, *Inorg. Chem.*, **1**, 204 (1962).

2.10. 2,3-DIMETHYL-2-BUTYLBORANE [THEXYLBORANE]

$$(H_3C)_2C{=}C(CH_3)_2 + H_3B{:}THF \xrightarrow[0^\circ]{THF} H{-}C(CH_3)_2{-}C(CH_3)_2{-}BH_2$$

Procedure by George Zweifel[1]

Procedure

A carefully dried 200-ml reaction flask, fitted with a septum inlet and a magnetic stirring bar, is attached to a mercury bubbler **(P2.9:N1)**. The flask is flushed with dry nitrogen and then maintained under a static pressure of nitrogen **(S9.1)**. In the flask is placed 50.0 ml of a 2.00*M* solution of borane-tetrahydrofuran **(P2.3) (P2.9:N8)**. The flask is immersed in an ice-salt bath, and the temperature is lowered to -10 to -15°. A hypodermic syringe is used to measure out 50 ml of a 2.00*M* solution of 2,3-dimethyl-2-butene **(N1)**, and the solution is then added slowly through the serum cap to the stirred borane reagent over a period of approximately 30 minutes, maintaining the temperature at approximately 0° **(N2)**. The reaction mixture is allowed to stand for 2 hours at 0° to complete the reaction. A clear solution of thexylborane, 1.00*M*, in tetrahydrofuran is thus obtained **(N3)**. It can be utilized *in situ* for hydroborations **(S3, P4.16, P4.17, P4.18)**. Alternatively, aliquots can be removed with a hypodermic syringe and utilized **(N4)**.

Notes

1. 2,3-Dimethyl-2-butene, bp 73.2°, n^{20}D 1.4122 **(S9.11)**.
2. Alternatively, 8.4 g of the pure olefin can be introduced to provide a more concentrated solution of the reagent.
3. Thexylborane is dimeric in tetrahydrofuran solution.[2] It can be isolated as a water-white liquid, mp -34.7 to -32.3°.[2]
4. At 25° it slowly undergoes migration of the boron atom from the tertiary position to the primary (3% in 8 days, 9% in 16 days).[1] Its stability at 0° has not been explored. In our studies we have found it convenient to synthesize

thexylborane fresh from standard solutions of 2,3-dimethyl-2-butene and borane-tetrahydrofuran.

[1]G. Zweifel and H. C. Brown, *J. Amer. Chem. Soc.*, **85**, 2066 (1963).
[2]H. C. Brown and G. J. Klender, *Inorg. Chem.*, **1**, 204 (1962).

2.11. 9-BORABICYCLO[3.3.1]NONANE[9-BBN]; *cis*-1,5-CYCLOOCTANEDIOL

$$\text{(1,5-cyclooctadiene)} + H_3B{:}THF \xrightarrow[0^\circ]{THF} \left[\begin{array}{c}1,4- \;+\\ 1,5-\end{array}\right] \xrightarrow[\Delta]{THF} \text{9-BBN}$$

$$\text{9-BBN (BH)} + C_2H_5OH \xrightarrow[25^\circ]{THF} \text{B-}OC_2H_5 + H_2$$

$$\text{B-}OC_2H_5 + 2\,H_2O_2 + NaOH \xrightarrow[40\text{-}50^\circ]{THF\text{-}EtOH} \textit{cis}\text{-1,5-cyclooctanediol (OH, HO)} + C_2H_5OH + NaB(OH)_4$$

Procedure by Evord F. Knights and Charles G. Scouten[1]

Procedure

A 1-ℓ single-necked flask fitted with a septum inlet (**N1**) and magnetic stirring bar is assembled with a 125-ml pressure-equalizing dropping funnel surmounting a short condenser connected by a ground joint to the flask (**F9.6**). The dropping funnel is capped by an adapter (**F9.2a**) which leads through a hypodermic needle to a mercury bubbler. The assembly is flushed with nitrogen introduced through the septum inlet (A, **F9.6**) and then maintained under a static pressure of the gas (**S9.1**, **F9.1**). The reaction flask is immersed in an ice-salt bath, and 150 ml of a 2.00*M* borane-tetrahydrofuran complex (**P2.3**)(**P2.9**:**N8** is introduced (0.300 mole BH_3) through the septum inlet on the flask (A) (via double-ended needle, **S9.2**).

Then 150 ml of a 2.00*M* solution of 1,5-cyclooctadiene (**N2**) is introduced into the dropping funnel and added dropwise to the rapidly stirred solution over 30 to 60 minutes maintaining the temperature at approximately 0°.

Finally, 10 ml of tetrahydrofuran is added through the dropping funnel to wash residual diene into the reaction flask. After the addition is complete, the cooling bath is replaced by a heating mantle, and the mixture is heated under gentle reflux for 1 hour (**N3**). This solution can be utilized to prepare (*a*) solid 9-BBN, (*b*) a standard solution of 9-BBN (0.50*M*), or (*c*) the oxidation product, *cis*-1,5-cyclooctanediol.

To obtain solid 9-BBN, the solution is slowly cooled first to room temperature and then to 0°. The 9-BBN precipitates. The supernatant liquid is removed with a double-ended needle, and the solid is washed with *ice-cold,* olefin-free pentane (2 × 30 ml). The flask and contents are heated at 50° under vacuum. There is obtained 22 g of solid 9-BBN (60% yield), mp 152-155° (sealed, evacuated capillary, **S9.10**).

To prepare the standard solution, 290 ml of tetrahydrofuran is added to the hot tetrahydrofuran solution (following the isomerization stage), and the total 600 ml is allowed to come to room temperature. With a stream of nitrogen passing through a hypodermic needle inserted into the septum inlet of the flask, the dropping funnel-condenser assembly (**N4**) is lifted up, and the single neck is closed with a glass stopper (**N5**). This procedure provides 600 ml of a 0.5*M* solution of 9-BBN in tetrahydrofuran (**N6**).

For oxidation to the diol, the solution, following the isomerization step, is brought to room temperature, and 23 g (29 ml, 0.50 mole) of ethanol is added (**N7**). After hydrogen evolution (**P2.1:N5**) has ceased, 100 ml of 3*M* sodium hydroxide is added rapidly.

The flask is immersed in an ice-water bath, and 67 ml of 30% hydrogen peroxide is added dropwise (exothermic), maintaining the temperature below 40-50° (**P2.1:N6, N7**). The reaction mixture is then stirred at 40-50° for an additional hour to ensure completion of the oxidation. Potassium carbonate is added to saturate the aqueous phase (**N8**). The tetrahydrofuran phase is dried over potassium carbonate, and the tetrahydrofuran is removed under vacuum. There is obtained an essentially quantitative yield of *cis*-1,5-cyclooctanediol. Vacuum distillation followed by recrystallization from ether yields *cis*-1,5-cyclooctanediol, mp 73.5-74.8° (lit.[2] mp 73.8-74.8°).

Notes

1. If the flask is to be used to store a standard solution for an appreciable time, the inlet should be protected by a Teflon stopcock (**S9.2, F9.6**).
2. 1,5-Cyclooctadiene, bp 45-48° (19 mm), n^{20}D 1.4938, was distilled from a small quantity of lithium aluminum hydride. A foam guard should be used. It is essential that the diene be added to the borane, and not vice versa.
3. The heating period is necessary to isomerize 1,4-bora moieties to the desired 1,5-.

4. This arrangement is primarily to minimize the number of openings in a flask to be used for storage of a standard solution. For the preparation of 9-BBN to be used directly for oxidation or hydroboration (**P4.4, P4.5, P4.6**) the more customary assembly (**P2.4**) is entirely satisfactory.
5. Alternatively, the solution could be made in an ordinary flask and transferred with a double-tipped needle to a storage flask (**S9.9**).
6. The concentration can be checked by hydrolysis (**S9.10**). Solutions of 9-BBN in tetrahydrofuran have been stored for over a year at room temperature without observable change. Such 0.5*M* solutions, as well as solid 9-BBN, are now available from the Aldrich Chemical Company.
7. Water is not used as in **P2.1**, since solid 9-BBN can precipitate from the concentrated solution, and the solid undergoes hydrolysis slowly.
8. Approximately 1 g of potassium carbonate per milliliter of water added is used. Thus this preparation involved 67 ml of base and 67 ml of peroxide, so that 134 g of potassium carbonate would be added. Excess potassium carbonate should not be added since it interferes with the stirring and separation.

[1] E. F. Knights, C. G. Scouten, and H. C. Brown, manuscript in preparation.
[2] A. C. Cope and L. L. Estes, Jr., *J. Amer. Chem. Soc.*, **72**, 1128 (1950).

2.12. 3,5-DIMETHYLBORINANE

$$+ \ H_3B{:}THF \xrightarrow[0\text{-}25^\circ]{THF} \xrightarrow[1\,hr,\,70^\circ]{\Delta}$$

B
H

Procedure by Ei-ichi Negishi[1]

Procedure

The usual hydroboration apparatus is assembled, dried, and flushed with nitrogen (**S9.1, F9.4**). In the flask is placed 100 ml (0.200 mole) of a 2.00*M* solution of borane-tetrahydrofuran complex (**P2.3**) (**P2.4:N3**). The flask is cooled in an ice-water bath and 19.2 g (0.200 mole) of 2,4-dimethyl-1,4-pentadiene (**N1**) is added dropwise over 0.5 hour, allowing the temperature to rise to 25°. The cold bath is removed, and the reaction mixture is heated under gentle reflux (~70°) for 1 hour (**N2**). Distillation (**S9.7, P2.7**) provides 19.2 g (87%) of 3,5-dimethylborinane, bp 79-82° (1 mm). For most applications the reagent can be utilized *in situ* (**P4.8**), or the solution can be stored and aliquots removed for various applications. In the latter case it is desirable that the reaction flask be fitted with an inlet covered with a rubber septum cap to facilitate removal of the aliquots by hypodermic syringe without the necessity of opening the apparatus (**P2.11:N1**). The same procedure can be utilized for the synthesis of 3,6-dimethylborepane[1] (**N3**).

Notes

1. 2,4-Dimethyl-1,4-pentadiene (n^{20}_{D} 1.4179) and 2,5-dimethyl-1,5-hexadiene (n^{20}_{D} 1.4290) were used as supplied by the Chemical Samples Company. VPC examination indicated the purities to be at least 99%. See **S9.11** for purification.
2. Methanolysis of aliquots and VPC analysis for the B-methoxy derivative indicated the presence of only 50% of the desired reagent immediately following the hydroboration stage. This rose to 94% after 1 hour of reflux.
3. 3,6-Dimethylborepane underwent decomposition upon attempted distillation at 0.1 mm.

[1] E. Negishi and H. C. Brown, *J. Amer. Chem. Soc.*, **95**, 6757 (1973).

3

HYDROBORATION WITH BORANE DERIVATIVES: SURVEY

The hydroboration reaction based on borane (**S1**) is so broadly applicable and so simple to apply that it would appear to constitute a model for an ideal reaction in synthetic chemistry. However, on occasion the polyfunctional nature of borane, BH_3, its relatively low selectivity, and its low steric requirements result in difficulties.

For example, the powerful directive influence of chlorine in allyl chloride leads to approximately a 50:50 distribution of the boron at the terminal and nonterminal positions.

$$\underset{\uparrow \atop 50\%}{CH_2}{=}\underset{\uparrow \atop 50\%}{CH}CH_2Cl$$

Hydroboration of *cis*-4-methyl-2-pentene results in little selectivity between the two positions.

$$(CH_3)_2CH\underset{\uparrow \atop 43\%}{CH}{=}\underset{\uparrow \atop 57\%}{CH}CH_3 \quad (cis)$$

The hydroboration of terminal acetylenes largely proceeds past the vinylborane stage.

$$RC{\equiv}CH \longrightarrow \underset{H}{\overset{R}{}}C{=}C\underset{B-}{\overset{H}{}} \longrightarrow RCH_2{-}CH(B{<})_2$$

Finally, the hydroboration of dienes is frequently complicated by the formation of cyclic intermediates.

$$H_2C(CH{=}CH_2)_2 \longrightarrow \begin{matrix} 62\%\ 1,4\text{-} \\ 38\%\ 1,5\text{-} \end{matrix}$$

Many of these difficulties can be circumvented by carrying out the hydroboration with borane intermediates.[1-3]

3.1. DISIAMYLBORANE

The reaction of 2-methyl-2-butene with borane proceeds rapidly to the dialkylborane stage, but only slowly beyond.

$$H_3B \xrightarrow[0^\circ,\ \text{fast}]{(CH_3)_2C{=}C(CH_3)H} (H{-}C(CH_3)_2{-}CH(CH_3){-})_2BH \xrightarrow[0^\circ,\ \text{slow}]{(CH_3)_2C{=}C(CH_3)H} (H{-}C(CH_3)_2{-}CH(CH_3){-})_3B$$

Consequently, this observation suggested that this monofunctional reagent might possess an enhanced selectivity over that of borane itself.[4]

Indeed, allyl chloride,[5] 1-hexene,[4] and styrene[4] all undergo hydroboration with disiamylborane (**P2.9**) to place 98-99% of the boron at the terminal position (**P4.1**).

$$\underset{98\%}{\overset{\uparrow}{H_2C{=}CH{-}CH_2Cl}} \qquad \underset{99\%}{\overset{\uparrow}{H_2C{=}CH{-}(CH_2)_3CH_3}} \qquad \underset{98\%}{\overset{\uparrow}{H_2C{=}CH{-}C_6H_5}}$$

[1] H. C. Brown, *Hydroboration,* W. A. Benjamin, New York, 1962.
[2] H. C. Brown, *Boranes in Organic Chemistry,* Cornell University Press, Ithaca, N.Y., 1972.
[3] G. M. L. Cragg, *Organoboranes in Organic Synthesis,* Marcel Dekker, New York, 1973.
[4] H. C. Brown and G. Zweifel, *J. Amer. Chem. Soc.,* 83, 1241 (1961).

The reagent also exhibits the desired stereoselectivity for the less hindered of the two positions of a double bond.[4,6]

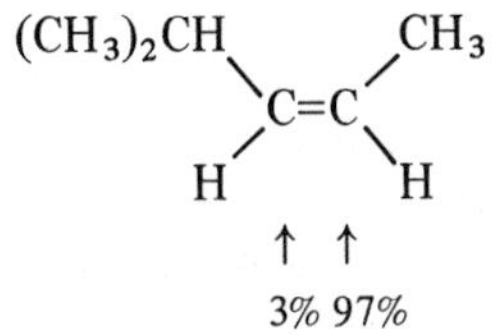

$(CH_3)_2CH$ CH_3
C=C
H H
↑ ↑
3% 97%

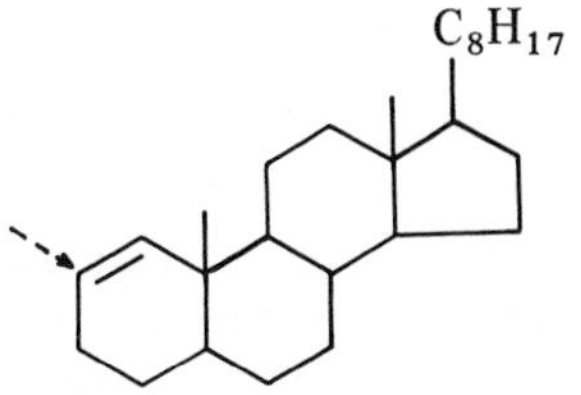

C_8H_{17}

Disiamylborane is also highly sensitive to the structure of the olefin. Thus terminal olefins, such as 1-hexene and 2-methyl-1-pentene, react much more rapidly than internal olefins, such as *trans*-2-hexene. *cis*-Alkenes react considerably faster than *trans*. Cyclopentene reacts approximately 100-fold faster than cyclohexene. Trisubstituted olefins, such as 2-methyl-2-butene, react comparatively slowly.[4,7]

These characteristics make possible the selective hydroboration of one olefin in a mixture, or more important, the selective hydroboration of a particular double bond in a structure containing two or more such bonds[4] (**P4.2**).

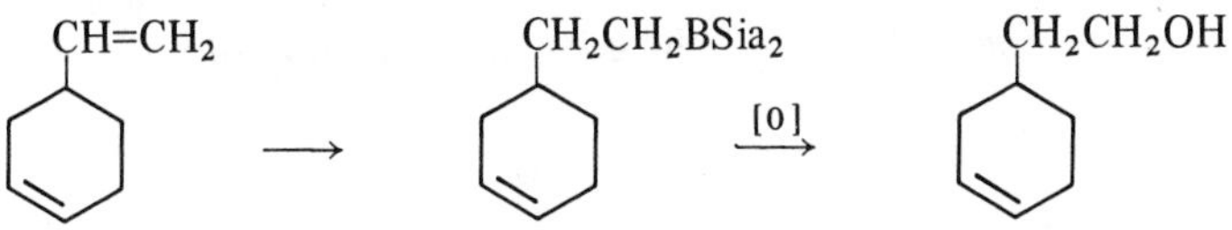

$CH{=}CH_2$ → $CH_2CH_2BSia_2$ $\xrightarrow{[O]}$ CH_2CH_2OH

The monofunctional nature of disiamylborane also circumvents the difficulty previously pointed out of accomplishing the simple dihydroboration of dienes.[8]

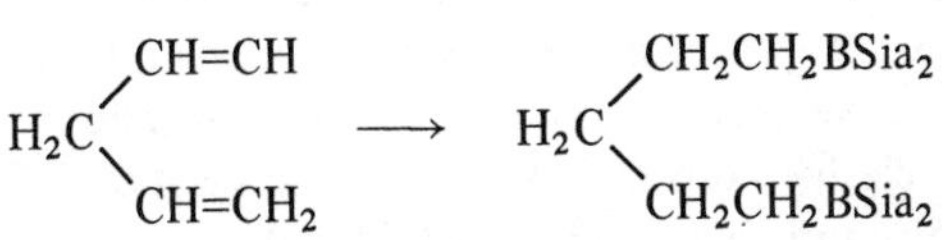

$H_2C(CH{=}CH)(CH{=}CH_2)$ → $H_2C(CH_2CH_2BSia_2)(CH_2CH_2BSia_2)$

Finally, disiamylborane permits the mononydroboration of acetylenes, making readily available the corresponding vinyl boron derivatives.[9] This development makes possible a simple anti-Markovnikov hydration of terminal acetylenes[9] (**P4.3**).

[5] H. C. Brown and K. A. Keblys, *J. Amer. Chem. Soc.*, **86**, 1791 (1964).
[6] F. Sondheimer and M. Nussim, *J. Org. Chem.*, **26**, 630 (1961).
[7] H. C. Brown and A. W. Moerikofer, *J. Amer. Chem. Soc.*, **83**, 3417 (1961).
[8] G. Zweifel, K. Nagase, and H. C. Brown, *J. Amer. Chem. Soc.*, **84**, 183 (1962).
[9] H. C. Brown and G. Zweifel, *J. Amer. Chem. Soc.*, **83**, 3834 (1961).

$$RC{\equiv}CH \xrightarrow{Sia_2BH} \underset{H}{\overset{R}{}}\!\!>C{=}C<\!\!\underset{BSia_2}{\overset{H}{}} \xrightarrow{[O]} RCH_2CHO$$

In many cases dicyclohexylborane (**P2.8**) can substitute for disiamylborane. On occasion, the lower steric requirements of dicyclohexylborane can offer advantages, as in the reaction with mercuric acetate to produce the desired vinylmercuric acetates.[10] With these reagents it is also possible to take advantage of powerful directive effects to achieve interesting syntheses.[11]

$$CH_3C{\equiv}CCO_2R \xrightarrow{(C_6H_{11})_2BH} \underset{H}{\overset{CH_3}{}}\!\!>C{=}C<\!\!\underset{B(C_6H_{11})_2}{\overset{CO_2R}{}} \xrightarrow{[O]} CH_3CH_2COCO_2R$$

These reagents also provide a simple route from diynes to *cis*-dienes.[12]

3.2. 9-BBN

The reaction of 1,5-cyclooctadiene with borane in tetrahydrofuran can be controlled to provide the bicyclic borane. 9-borabicyclo[3.3.1]nonane[13] (**P2.11**).

$$\text{1,5-cyclooctadiene} \xrightarrow{H_3B:THF} \text{9-BBN (H–B bicyclic)}$$

This reagent, termed 9-BBN for convenience, possesses remarkable thermal stability. Moreover, it possesses remarkable air stability for a dialkylborane, permitting it to be weighed out and transferred with approximately the same precautions and relative convenience accompanying the utilization of sodium borohydride and lithium aluminum hydride. Furthermore, its present commercial availability[14] should facilitate many synthetic applications of borane chemistry.

The rate of reaction of 9-BBN with olefins is considerably slower than that of disiamylborane. However, terminal olefins react readily at 25°. Less reactive internal olefins require either an excess of reagent or overnight reaction times or both for complete hydroboration.[13]

[10] R. C. Larock and H. C. Brown, *J. Organomet. Chem.*, **36**, 1 (1972).
[11] J. Plamondon, J. T. Snow, and G. Zweifel, *Organomet. Chem. Syn.*, **1**, 249 (1971).
[12] G. Zweifel and N. L. Polston, *J. Amer. Chem. Soc.*, **92**, 4068 (1970).
[13] E. F. Knights, C. G. Scouten, and H. C. Brown, *J. Amer. Chem. Soc.*, in press.
[14] Produced by Aldrich-Boranes, Inc., a subsidiary of Aldrich Chemical Company, Inc.

Fortunately, both 9-BBN and the B—R—9-BBN molecules exhibit remarkable thermal stabilities for organoboranes. Consequently, the hydroborations can be carried out without difficulty in refluxing tetrahydrofuran, utilizing the theoretical quantity of 9-BBN. Under these conditions, almost all olefins are quantitatively converted to the B—R—9-BBN derivatives within 1 hour (**P4.4**). Even the most resistant olefin examined, 2,3-dimethyl-2-butene, which fails to react with disiamylborane, undergoes complete hydroboration under these conditions in 8 hours.

9-BBN exhibits a remarkable regioselectivity in its hydroborations. Terminal olefins react to place the boron at the terminal position with a selectivity of at least 99.9%.

$$RCH{=}CH_2 \xrightarrow{\text{9-BBN}} RCH_2CH_2\text{-B} \xrightarrow{[O]} RCH_2CH_2OH \quad (>99.9\%)$$

Even more surprising is the regioselectivity exhibited on *cis*-4-methyl-2-pentene.[13]

$$(CH_3)_2CH(H)C{=}C(H)CH_3 \xrightarrow{\text{9-BBN}} (CH_3)_2CHCH_2CH(\text{B})CH_3 \quad 99.8\%$$

(Hydroboration of the *trans* isomer is much less regioselective.)

The powerful regioselectivity of the reagent can even overcome the major directive influence of halogen substituents (**P4.6, P8.2**).

$$H_2C{=}CHCH(Cl)CH_2Cl \xrightarrow{\text{9-BBN}} \text{BCH}_2CH_2CH(Cl)CH_2Cl \xrightarrow{OH^-} \triangleright\text{-}CH_2Cl$$

Acetylenes are readily converted to 1,1-dibora derivatives (**P4.6**).

$$RC{\equiv}CH + 2\ 9\text{-BBN} \longrightarrow RCH_2CH(B)_2$$

These are finding interesting new synthetic applications **(S7)**.

Many of the new reactions of organoboranes utilize only one of the three alkyl groups in the trialkylborane molecule **(S7)**. For many of these reactions the use of the B–R–9-BBN derivatives solves the problem–only the B–R groups participate in the desired reaction **(S7)**.

9-BBN; CO, $LiAlH(OCH_3)_3$; [O]; CHO

This characteristic of the 9-BBN system makes it highly desirable to be able to introduce groups in the 9-position which cannot be obtained via hydroboration. Such derivatives are now readily available through the reaction of B-methoxy-9-BBN with lithium alkyls or aryls[15] **(P4.7)**.

$$BOCH_3 + LiPh \xrightarrow{\text{Pentane}} BPh + LiOCH_3\downarrow$$

The reaction can be extended to other borinic acid derivatives.[15] For example, B-*t*-butyl-3,5-dimethylborinane **(P4.8)** is valuable for the transfer of the *t*-butyl group in certain 1,4-conjugate additions **(S7, P8.24)**.

B, OCH_3; *t*-BuLi, pentane; B, $H_3C-C-CH_3$, CH_3; O; O, C, CH_3, CH_3, CH_3

[15]G. W. Kramer and H. C. Brown, *J. Organomet. Chem.*, 73, 1 (1974).

3.3. DIISOPINOCAMPHEYLBORANE

The hydroboration of α-pinene proceeds readily to the formation of diisopinocampheylborane (**P2.1**). Since α-pinene is available from natural sources in both optically active forms, (+)- and (–), this reaction makes available an optically active dialkylborane in both optically active forms.[16,17] This reagent, **IPC_2BH**, exhibits a remarkable ability to achieve asymmetric syntheses with certain olefins.

For example, *cis*-2-butene reacts with the reagent from α-pinene $[\alpha]^{20}D$ +47.6°, to produce an organoborane which, when oxidized with alkaline hydrogen peroxide, produces (–)-2-butanol, $[\alpha]^{20}D$ −11.8°[14] (**P4.9**).

$$IPC_2BH \xrightarrow{CHCH_3=CHCH_3} IPC_2B-\overset{*}{C}H(CH_2CH_3)CH_3 \xrightarrow{[O]} HO\overset{*}{-}C(CH_3)(CH_2CH_3)-H$$

This rotation corresponds to an optical purity of 87%. Use of (–)-α-pinene produces (+)-2-butanol, $[\alpha]^{20}D$ +11.8°.

Similarly, *cis*-3-hexene and norbornene have been converted into the corresponding alcohols with high optical purities.[17]

Treatment of a racemic mixture of 3-methylcyclopentene with the reagent results in a selective reaction of one of the enantiomers. 3-Methylcyclopentene was recovered in an optical purity as high as 65%.[17] IPC_2BH has recently been utilized to achieve an asymmetric synthesis of a prostaglandin intermediate.[18]

$CH_2CO_2CH_3$ (cyclopentadiene) $\xrightarrow{(+)\text{-}IPC_2BH}$ $\xrightarrow{[O]}$ HO, $CH_2CO_2CH_3$ (cyclopentene)

$[\alpha]_D$ −136°
92% optically pure

The reagent also has interesting possibilities for asymmetric reductions.[2]

[16] G. Zweifel and H. C. Brown, *J. Amer. Chem. Soc.*, **86**, 393 (1964).
[17] H. C. Brown, N. R. Ayyangar, and G. Zweifel, *J. Amer. Chem. Soc.*, **86**, 397 (1964).
[18] J. J. Partridge, N. K. Chadha, and M. R. Uskoković, *J. Amer. Chem. Soc.*, **95**, 7171 (1973).

3.4. CATECHOLBORANE

Alcohols, such as methanol, react readily with borane in tetrahydrofuran to form dialkoxyboranes.[19]

$$2\ CH_3OH + BH_3 \xrightarrow{THF} (CH_3O)_2BH + 2\ H_2$$

Glycols undergo a similar reaction.[20] However, these derivatives are relatively

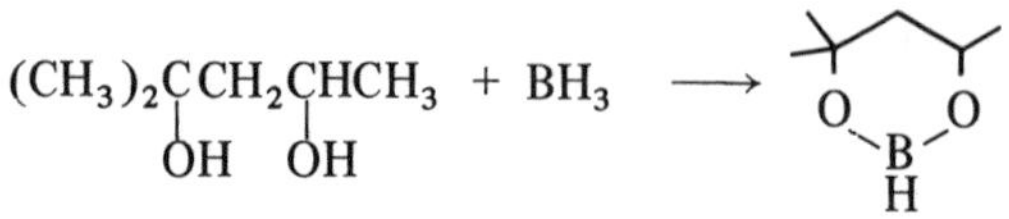

$$(CH_3)_2\underset{OH}{C}CH_2\underset{OH}{C}HCH_3 + BH_3 \longrightarrow$$

poor hydroborating agents.

Catechol undergoes a similar reaction[21] (**P4.10**).

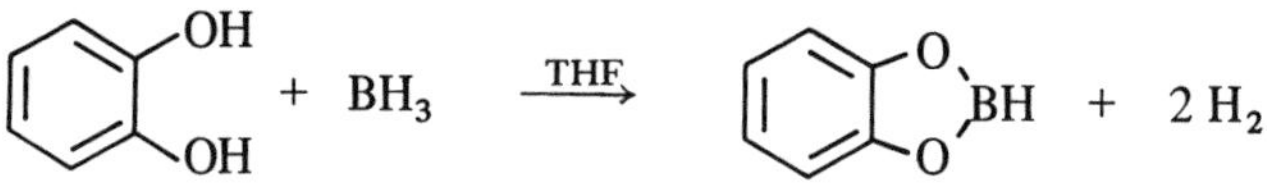

$$+ BH_3 \xrightarrow{THF} \quad BH + 2\ H_2$$

The product, 1,3,2-benzodioxaborole ("catecholborane"), is a considerably better hydroborating agent than the alkoxy derivatives. It readily hydroborates olefins at 100°.[21]

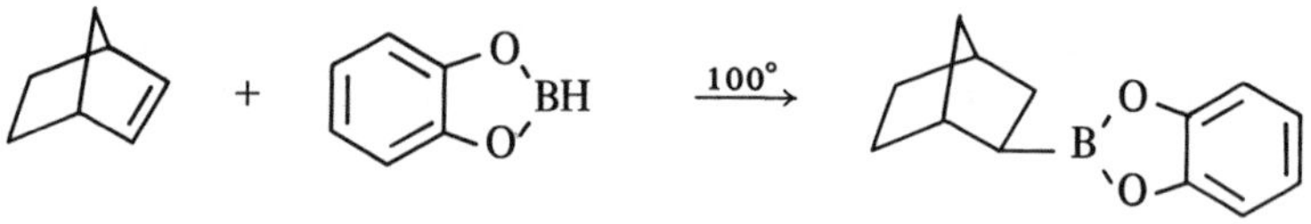

$$+ \quad BH \xrightarrow{100°}$$

The reaction with acetylenes is more facile, proceeding at a satisfactory rate at 70°[22] (**P4.11**).

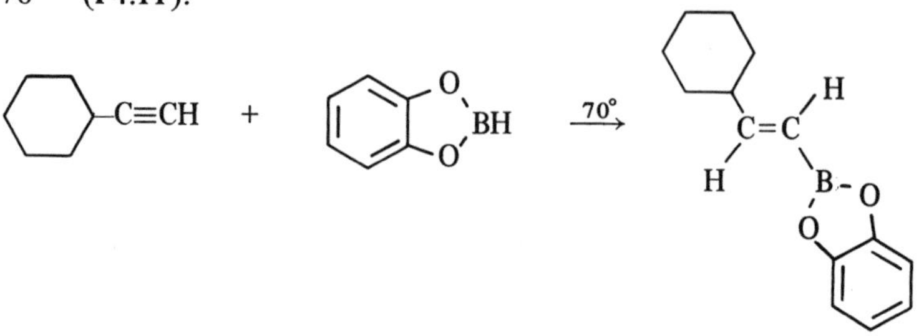

$$C{\equiv}CH + \quad BH \xrightarrow{70°} C{=}C$$

[19]Unpublished research with D. B. Bigley.
[20]W. G. Woods and P. L. Strong, *J. Amer. Chem. Soc.*, 88, 4667 (1966).
[21]H. C. Brown and S. K. Gupta, *J. Amer. Chem. Soc.*, 93, 1816 (1971).

The reagent reacts to place the boron atom preferentially at the least hindered position of the triple bond.[22]

$$(CH_3)_3CC{\equiv}CCH_3 + \text{catecholborane (C}_6\text{H}_4\text{O}_2\text{BH)} \rightarrow (H_3C)_3C(H)C{=}C(CH_3)B(O_2C_6H_4) \xrightarrow{[O]} (CH_3)_3CCH_2COCH_3$$

Oxidation then gives the corresponding carbonyl derivative.

A significant advantage of these catecholborane derivatives over the corresponding dialkylborane hydroboration products is their relatively simple conversion into the boronic acids by hydrolysis.

$$C_6H_4O_2BR + 2H_2O \longrightarrow RB(OH)_2 + C_6H_4(OH)_2$$

3.5. CHLOROBORANES

Dichloroborane, Cl_2BH, and monochloroborane, $ClBH_2$, appear to be unstable substances which disproportionate into diborane and boron trichloride spontaneously. However, in ether solvents diborane reacts readily with boron trichloride to form the corresponding etherates.[23]

$$B_2H_6 + 4\,Cl_3B{:}OEt_2 + 2\,OEt_2 \xrightarrow{EE} 6\,Cl_2BH{:}OEt_2$$

$$B_2H_6 + Cl_3B{:}OEt_2 + 2\,OEt_2 \xrightarrow{EE} 3\,ClBH_2{:}OEt_2$$

Alternatively, hydrogen chloride will react with borane-tetrahydrofuran to produce the corresponding complexes.[24]

$$H_3B{:}THF + 2\,HCl \xrightarrow{THF} Cl_2BH{:}THF + 2\,H_2$$

$$H_3B{:}THF + HCl \xrightarrow{THF} ClBH_2{:}THF + H_2.$$

[22] H. C. Brown and S. K. Gupta, *J. Amer. Chem. Soc.*, 94, 4370 (1972).
[23] H. C. Brown and P. A. Tierney, *J. Inorg. Nucl. Chem.*, 9, 51 (1959).
[24] G. Zweifel, *J. Organomet. Chem.*, 9, 215 (1967).

The difficulties involved in measuring and introducing the stoichiometric quantities of gaseous diborane handicapped the exploration of the ethyl etherates. The synthesis of the tetrahydrofuranates was far simpler. Unfortunately, the latter complexes are too stable, and the hydroboration reactions proved to be slow and incomplete.[24,25]

The ethyl etherates proved more favorable.[26] The reaction of boron trichloride in ether with lithium borohydride in ether provided a simple route to the chloroborane ethyl etherates[23] **(P4.12)**.

$$LiBH_4 + BCl_3 + 2\,Et_2O \xrightarrow{EE} LiCl\downarrow + 2\,ClBH_2{:}OEt_2$$

$$LiBH_4 + 3\,BCl_3 + 4\,Et_2O \xrightarrow{EE} LiCl\downarrow + 4\,Cl_2BH{:}OEt_2$$

Chloroborane ethyl etherate reacts readily at 0° with olefins[26] **(P4.14)**.

$$2\ \text{(cyclopentene)} + H_2BCl{:}OEt_2 \xrightarrow[0^\circ,\ 1\ hr]{EE} (\text{cyclopentyl})_2BCl + Et_2O$$

It exhibits a very powerful directive effect in the hydroboration of olefins, far greater than that of borane itself, and comparable to 9-BBN in many cases.[27]

$n\text{-}C_4H_9CH{=}CH_2$	$C_2H_5C(CH_3){=}CH_2$	$(CH_3)_2C{=}CHCH_3$
↑	↑	↑
> 99.5%	>99.9%	99.7%

Acetylenes also can be converted directly to the corresponding divinylborinic acid derivatives[28] **(P4.15)**.

$$2\,RC{\equiv}CR + H_2BCl{:}OEt_2 \xrightarrow{EE} (RHC{=}CR)_2BCl$$

These can be utilized directly in the Zweifel diene synthesis[29] **(P8.19)**.

[25] D. J. Pasto and P. Balasubramaniyan, *J. Amer. Chem. Soc.*, 89, 295 (1967).
[26] H. C. Brown and N. Ravindran, *J. Amer. Chem. Soc.*, 94, 2112 (1972).
[27] H. C. Brown and N. Ravindran, *J. Org. Chem.*, 38, 182 (1973).
[28] H. C. Brown and N. Ravindran, *J. Org. Chem.*, 38, 1617 (1973).
[29] G. Zweifel, N. L. Polston, and C. C. Whitney, *J. Amer. Chem. Soc.*, 90, 6243 (1968).

The dichloroborane-ethyl etherate fails to react spontaneously with olefins or acetylenes at any convenient rate.[30] If a long reaction time is utilized, the product is complex, evidently the result of disproportionation. Fortunately, if the 1:1 compound is added to a mixture of the olefin (or acetylene) and boron chloride in pentane at 0°, a rapid reaction occurs leading directly to the desired monoalkyl (or monovinyl) boron dichloride[30] (**P4.13**).

$$\text{(1-methylcyclopentene)} + Cl_2BH{:}OEt_2 + BCl_3 \xrightarrow[0^\circ]{\text{pentane}} \text{(trans-2-methylcyclopentyl)}BCl_2 + Cl_3B{:}OEt_2\downarrow$$

These derivatives, R_2BCl and $RBCl_2$, are much more powerful electrophiles than the corresponding R_3B compounds. They facilitate many reactions which proceed sluggishly or not at all with the parent organoborane (**S5, P6.15, S7, P8.16**).

More recently we have established that the addition compounds, $H_3B{:}S(CH_3)_2$ and $Cl_3B{:}S(CH_3)_2$, can be mixed in the proper proportions to yield the corresponding mono- and dichloroborane adducts.[31]

$$H_3B{:}S(CH_3)_2 + 2Cl_3B{:}S(CH_3)_2 \longrightarrow 3\ Cl_2BH{:}S(CH_3)_2$$

$$2H_3B{:}S(CH_3)_2 + Cl_3B{:}S(CH_3)_2 \longrightarrow 3\ ClBH_2{:}S(CH_3)_2$$

These derivatives apparently can be utilized in place of the ethyl etherates. The marked improvement in the ease of synthesizing these reagents should greatly facilitate their application in synthesis.

3.6. THEXYLBORANE

Thexylborane is readily prepared by treating 1 mole of borane in tetrahydrofuran with 1 mole of 2,3-dimethyl-2-butene[32] (**P2.10**).

$$(H_3C)_2C{=}C(CH_3)_2 + BH_3 \longrightarrow H{-}C(CH_3)_2{-}C(CH_3)_2{-}BH_2$$

It is the most readily available of the monoalkylboranes. It has proven to be exceptionally valuable for the cyclic hydroboration of dienes and numerous other applications.[33]

[30] H. C. Brown and N. Ravindran, *J. Amer. Chem. Soc.*, 95, 2396 (1973).
[31] Research in progress with N. Ravindran.
[32] G. Zweifel and H. C. Brown, *J. Amer. Chem. Soc.*, 85, 2066 (1963).
[33] E. Negishi and H. C. Brown, *Synthesis*, 77(1974).

The reaction of borane, a trifunctional reagent, with bifunctional dienes usually leads to the formation of polymers.[34] Utilization of the bifunctional reagent, thexylborane, greatly simplifies the synthesis of cyclic derivatives.[35]

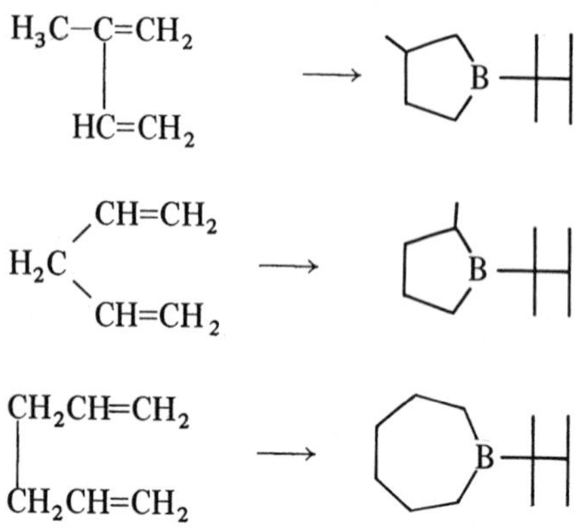

These derivatives are readily converted into cyclic ketones via the carbonylation (**S7.4, P8.6**) and cyanidation reactions (**S7.7, P8.8**).

The utility of this approach to cyclic hydroboration is indicated by the stereospecific conversion of D-(+)-limonene into the cyclic boron derivative (**P4.16**) readily converted into the *cis* diol or into pure D-(−)-carvomenthol.[36]

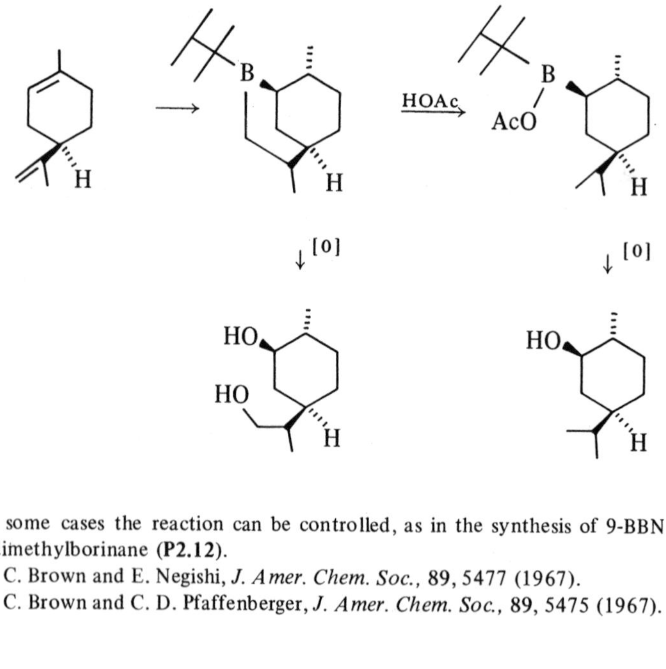

[34] In some cases the reaction can be controlled, as in the synthesis of 9-BBN (**P2.11**) and 3,5-dimethylborinane (**P2.12**).

[35] H. C. Brown and E. Negishi, *J. Amer. Chem. Soc.*, 89, 5477 (1967).

[36] H. C. Brown and C. D. Pfaffenberger, *J. Amer. Chem. Soc.*, 89, 5475 (1967).

This bicyclic boron intermediate has been converted into a trialkylborohydride which achieves superior stereospecificity in the reduction of a prostaglandin intermediate.[37]

Simple 1-alkenes react readily with thexylborane to produce the corresponding thexyldialkylboranes.[32]

$$\text{Thx–BH}_2 + 2\,\text{RCH=CH}_2 \longrightarrow \text{Thx–B(CH}_2\text{CH}_2\text{R)}_2$$

Treatment of thexylborane with one molar equivalent of such a 1-alkene fails to achieve the monoalkylation of thexylborane–a mixutre of products is obtained.[38] On the other hand, the reaction of more hindered olefins, such as 2-methyl-1-propene, cyclopentene, norbornene, and 3-methyl-2-butene, can be controlled to give excellent yields of the thexylmonoalkylboranes[38] **(P4.17)**.

$$\text{Thx–BH}_2 + \text{cyclohexene} \longrightarrow \text{Thx–B(H)(C}_6\text{H}_{11})$$

These monoalkyl derivatives can be hydroborated further with olefins of not too large steric requirements[39] **(P4.18)**.

$$\text{Thx–B(H)(C}_6\text{H}_{11}) + \text{CH}_2\text{=CHCH}_2\text{CO}_2\text{R} \longrightarrow \text{Thx–B(C}_6\text{H}_{11})\text{CH}_2\text{CH}_2\text{CH}_2\text{CO}_2\text{R}$$

A valuable reaction of the monoalkyl derivatives is their transformation with triethylamine into the monoalkylborane aminates **(P4.17)**.

$$\text{Thx–B(H)(C}_6\text{H}_{11}) + \text{Et}_3\text{N} \longrightarrow \text{C}_6\text{H}_{11}\text{–BH}_2\text{:NEt}_3 + (\text{CH}_3)_2\text{C=C(CH}_3)_2$$

[37] E. J. Corey and R. K. Varma, *J. Amer. Chem. Soc.*, **93**, 7319 (1971).
[38] E. Negishi, J.-J. Katz, and H. C. Brown, manuscript in preparation.
[39] E. Negishi, J.-J. Katz, and H. C. Brown, manuscript in preparation.

This reaction provides the first simple, general synthesis of monoalkylboranes.

The thexyldialkylboranes are valuable for the synthesis of ketones via the carbonylation (**S7.4**) and cyanidation reactions (**S7.7**) and the synthesis of tertiary alcohols via the α-bromination reaction (**S7.10**). For a fuller discussion of the remarkable versatility of thexylborane, the recent review should be examined.[33]

3.7. CONCLUSION

It is evident that to take full advantage of the characteristics of the hydroboration reaction and the remarkable versatility of the organoborane chemistry[2,3] (**S5** and **S7**), it is essential that the chemist become skilled in the selection and utilization of the particular borane derivative most advantageous for the chemical transformation required. It is hoped that this brief review will assist the organic chemist interested in synthesis toward that objective.

4

HYDROBORATION WITH BORANE DERIVATIVES: PROCEDURES

4.1. ETHYL 4-HYDROXYBUTYRATE

$$Sia_2BH + H_2C{=}CHCH_2CO_2Et \xrightarrow[0^\circ]{THF} Sia_2B(CH_2)_3CO_2Et$$

$$Sia_2B(CH_2)_3CO_2Et + 3\,H_2O_2 + NaOH \xrightarrow[25^\circ]{THF} HO(CH_2)_3CO_2Et + 2\,SiaOH + NaB(OH)_4$$

Procedure by Kestutis A. Keblys[1]

Procedure

Disiamylborane, 0.100 mole, is prepared as described (**P2.9**), utilizing a 500-ml flask to avoid transfer of the solution. A static nitrogen atmosphere is maintained until after the oxidation (**S9.1**). The flask is immersed in an ice bath, and 11.4 g (0.100 mole) of ethyl 3-butenoate (**N1**) is added to the stirred solution over a period of 0.5 hour. After another period of 0.5 hour, 3 ml of water is added to destroy residual hydride. The reaction mixture is immersed in a cold water bath and oxidized by the slow concurrent addition (**P2.1:N6**) of 33 ml of 3*M* sodium hydroxide and 33 ml of 30% hydrogen peroxide (**N2**), maintaining the temperature at 20-25°. The oxidation mixture is diluted with 50 ml of ether, and the layers are separated. The aqueous layer is extracted with ethyl ether (2 X 50 ml). The organic layers are combined and washed successively with 20 ml of 1% sodium carbonate, 20 ml of saturated sodium sulfite, and 15 ml of saturated sodium chloride. The solution is dried over magnesium sulfate. The solvent and most of the siamyl alcohol is removed by evaporation in vacuo at room temperature. The residue is distilled rapidly at 0.1 mm into an ice-cooled receiver. Redistillation of the flash distillate gives 10.0 g, a 76% yield of ethyl 4-hydroxybutyrate, collected at 43-44° at 0.15 mm, n^{20}D 1.4294 **(N4)**. VPC analysis indicated the isomeric purity to be at least 98% **(N5)**.

Notes

1. Prepared by the esterification of 3-butenoic acid, bp 124° (750 mm), n^{20}D 1.4103 **(S9.11)**.
2. The alkali and hydrogen peroxide are added concurrently in order to minimize hydrolysis of the ester.
3. Ethyl 4-hydroxybutyrate must be treated gently since it is readily converted into γ-butyrolactone and ethanol.[2]
4. The same procedure has been utilized for the conversion of ethyl 4-pentenoate into ethyl 5-hydroxyvalerate (78% yield) and ethyl 10-undecenoate into ethyl 11-hydroxyundecanoate (81% yield).[1]

Disiamylborane has also been used to convert the free acid, 10-undecenoic acid, into 11-hydroxyundecanoic acid (82% yield).[3] In this case, 2 moles of disiamylborane per mole of acid must be used.

$$2\,Sia_2BH + H_2C{=}CH(CH_2)_8CO_2H \longrightarrow Sia_2B(CH_2)_{10}CO_2BSia_2 + H_2$$

$$Sia_2B(CH_2)_{10}CO_2BSia_2 + 5\,H_2O_2 + 3\,NaOH \longrightarrow HO(CH_2)_{10}CO_2Na + 4\,SiaOH + 2\,NaB(OH)_4$$

5. Hydroboration with borane gives approximately 20% of the 3-hydroxy isomer.[1]

[1] H. C. Brown and K. A. Keblys, *J. Amer. Chem. Soc.*, **86**, 1795 (1964).
[2] H. C. Brown and K. A. Keblys, *J. Org. Chem.*, **31**, 485 (1966).
[3] H. C. Brown and D. B. Bigley, *J. Amer. Chem. Soc.*, **83**, 486 (1961).

4.2. 2-(4-CYCLOHEXENYL)ETHANOL

$$\text{(4-cyclohexenyl)}CH{=}CH_2 + Sia_2BH \xrightarrow[0^\circ]{THF} \text{(4-cyclohexenyl)}CH_2CH_2BSia_2$$

$CH_2CH_2BSia_2$ (cyclohexenyl) + 3 H_2O_2 + NaOH $\xrightarrow[40\text{-}50^\circ]{\text{THF}}$ CH_2CH_2OH (cyclohexenyl)

+ 2 SiaOH + $NaB(OH)_4$

Procedure by George Zweifel[1]

Procedure

A typical hydroboration setup is assembled (**P2.5**), utilizing a 500-ml flask. The system is flushed with nitrogen, and a static pressure of nitrogen is maintained until after the oxidation (**S9.1**). In the flask is placed 16.2 g (0.150 mole) of 4-vinylcyclohexene (**N1**) in 30 ml of tetrahydrofuran. The flask is immersed in an ice bath. In the dropping funnel is placed 165 ml (0.165 mole) of a 1.0*M* solution of disiamylborane (**N2**) in tetrahydrofuran (**P2.9**), transferred either by a hypodermic syringe or a double-ended stainless-steel needle (**S9.2**). The disiamylborane solution is added dropwise to the magnetically stirred diene solution over a period of 1 hour (**N3**). After stirring for an additional hour at 0°, 10 ml of water is added to destroy residual hydride (~400 ml H_2 evolved), and the cold bath is removed. The alkali, 55 ml of 3*M* sodium hydroxide, is added rapidly, followed by the slow dropwise addition of 55 ml of 30% hydrogen peroxide at a rate such that the temperature does not rise above 50° (**P2.1;N6**). The aqueous phase is saturated with sodium chloride. The tetrahydrofuran phase is separated and dried over magnesium sulfate. Distillation gives 13.7 g (72% yield) of 2-(4-cyclohexenyl)ethanol, bp 86-87° at 6 mm, n^{20}D 1.4834 (lit.[2] bp 104.5° at 16 mm, n^{20}D 1.4832).

Notes

1. Commercial, 99%, bp 130° (745 mm), n^{20}D 1.4637 (**S9.11**).
2. The use of a 10% excess is not really necessary if the apparatus and solvents have been thoroughly dried.
3. It is generally preferable to add the disiamylborane to the diene in order to achieve the selective hydroboration of the more reactive double bond. In the present case, the vinyl double bond is so much more reactive than the cyclohexene double bond that no difficulty would be anticipated in the simpler procedure of adding the diene directly to the disiamylborane solution at 0° (see **P4.3**).

[1] H. C. Brown and G. Zweifel, *J. Amer. Chem. Soc.*, **83**, 1241 (1961).
[2] J. J. Bost, R. E. Kepner, and A. B. Webb, *J. Org. Chem.*, **22**, 51 (1957).

4.3. OCTANAL

$$CH_3(CH_2)_5C{\equiv}CH + Sia_2BH \xrightarrow[0°]{THF} \underset{H}{\overset{CH_3(CH_2)_5}{}}\!\!>C{=}C<\!\!\underset{BSia_2}{\overset{H}{}}$$

$$\underset{H}{\overset{CH_3(CH_2)_5}{}}\!\!>C{=}C<\!\!\underset{BSia_2}{\overset{H}{}} + 3\,H_2O_2 + NaOH \xrightarrow[0°,\ pH8]{THF} CH_3(CH_2)_6CHO + 2\,SiaOH + NaB(OH)_4$$

Procedure by George Zweifel[1]

Procedure

Disiamylborane (0.200 mole) is prepared in a 500-ml flask (**P2.9**). A static pressure of nitrogen is maintained until after the oxidation (**S9.1**). The flask is immersed in an ice-salt bath and the temperature lowered to ~−10°. A solution of 22.0 g (0.200 mole) of 1-octyne (**N1**) in 20 ml of tetrahydrofuran is added as rapidly as possible while maintaining the temperature below 10°. The cold bath is removed and the reaction mixture allowed to come to room temperature for 1 hour to complete the hydroboration. The reaction mixture is cooled to 0° and the product oxidized at 0° by the addition of 150 ml of a 15% solution of hydrogen peroxide (**N2**), maintaining the pH of the solution at approximately pH 8 by the controlled addition of 67 ml of 3*M* sodium hydroxide (**N3**). After oxidation, the reaction mixture is brought to the neutral point and steam distilled. The distillate is extracted with ether, and the ether extract is dried over anhydrous magnesium sulfate (**N4**). Distillation provides 18 g (70% yield) of *n*-octaldehyde, bp 83-85° at 33 mm, n^{20}D 1.4217 (lit.[3] bp 171-173°, n^{20}D 1.4216).

Notes

1. Commercial 1-octyne was distilled from sodium borohydride to remove peroxides (bp 125° (745 mm), n^{20}D 1.4159).
2. The commercial 30% material was diluted with water to provide greater control over the vigor of the reaction and the temperature.
3. The usual procedure of adding all of the alkali initially results in more alkaline conditions and loss of aldehyde by condensation. Alternatively, it is possible to use a buffer to achieve oxidation to aldehyde.[2]
4. An alternative procedure to recover aldehyde from such oxidation mixtures has been developed (**P8.5**).

[1]H. C. Brown and G. Zweifel, *J. Amer. Chem. Soc.*, **83**, 3834 (1961).
[2]H. C. Brown and R. A. Coleman, *J. Amer. Chem. Soc.*, **91**, 4606 (1969).
[3]C. D. Harries and K. Oppenheim, *Chem. Zentr.*, **87**, 993 (1916).

4.4. B-(*TRANS*-2-METHYLCYCLOPENTYL)-9-BORABICYCLO[3.3.1] NONANE

+ 9-BBN —THF, reflux, 1 hr→ B

Procedure by Evord F. Knights and Charles G. Scouten[1]

Procedure

The usual apparatus (**S9.1**, **P2.5**) is assembled, utilizing a 300-ml flask. The system is flushed with nitrogen. Solid 9-BBN (**P2.11** or Aldrich) is introduced into the flask, 24.4 g (0.200 mole) (**N1**). The system is again flushed with nitrogen, and a static pressure of nitrogen is maintained. Tetrahydrofuran, 40 ml, is added. Then 16.4 g (0.200 mole) of 1-methylcyclopentene (**N2**) in 20 ml of tetrahydrofuran is introduced, and the reaction mixture is refluxed for 1 hour. On cooling, there is present in the flask approximately 100 ml of a 2.0*M* solution of B-(*trans*-2-methylcyclopentyl)-9-BBN (**N3** and **N4**).

Notes

1. Alternatively, the commercially available 0.5*M* solution of 9-BBN (Aldrich Chemical Company) can be used. This will provide a product which is more dilute, ~0.45*M*, following the introduction of the 1-methylcyclopentene as the neat hydrocarbon.
2. 1-Methylcyclopentene, bp 75-76° at 745 mm, n^{20}D 1.4326, was prepared by the dehydration of 1-methylcyclopentanol (with iodine) prepared from cyclopentanone and methylmagnesium iodide. It can be dried as in **S9.11**.
3. For most purposes the organoborane may be used without isolation; however, it may be readily isolated by removal of the THF and vacuum distillation under nitrogen (**S9.7**), b.p. 75-76° at 0.1 mm, yield 36 g (88%).
4. The product can be oxidized in the usual manner by adding 67 ml of 3*M* sodium hydroxide, followed by the slow dropwise addition of 67 ml of 30% hydrogen peroxide (exothermic! **P4.2**). Saturation of the aqueous phase with potassium carbonate will give a dry tetrahydrofuran solution containing an almost quantitative yield of *trans*-2-methylcyclopentanol, bp 150-151° at 740 mm, n^{20}D 1.4505, and *cis*-1,5-cyclooctanediol, mp 73.5-74.8°.

The product can be utilized to transfer the *trans*-2-methylcyclopentyl group into the α-position of esters, ketones, nitriles, and so on (**S7.9, P8.11**). It can also be used to prepare pure *trans*-2-methylcyclopentanecarboxaldehyde (**S7.5, P8.5**).

B → CHO

[1] E. F. Knights, C. G. Scouten, and H. C. Brown, *J. Amer. Chem. Soc.*, in press.

4.5. B-[2-(4-CYCLOHEXENYL)ETHYL]-9-BORABICYCLO[3.3.1] NONANE

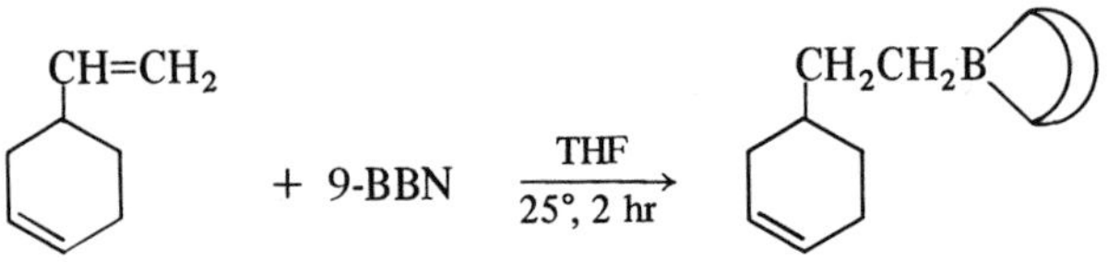

$CH=CH_2$ + 9-BBN $\xrightarrow[25°, 2 hr]{THF}$ CH_2CH_2B

Procedure by Evord F. Knights and Randolph A. Coleman[1]

Procedure

A dry 500-ml flask equipped with a septum inlet, dropping funnel, condenser, magnetic stirrer, and mercury bubbler is assembled (**S9.1**). The system is flushed with nitrogen and maintained under a static pressure of the gas. Into the flask is placed 10.8 g (0.100 mole) of 4-vinylcyclohexene (**P4.2:N1**). Then 200 ml of a 0.50*M* solution of 9-BBN in tetrahydrofuran (**N1**) is added over 0.5 hour. The reaction mixture is stirred for 1.5 hours (**N2**). There is obtained an essentially quantitative yield of the 9-BBN derivative (**N3** and **N4**).

Notes

1. Available from Aldrich Chemical Company. Alternatively, solid 9-BBN (**P2.11** or from Aldrich) can be dissolved in tetrahydrofuran, as described in **P4.4**.
2. 9-BBN exhibits high regiospecificity and high selectivity comparable to those of disiamylborane. Thus it can be used to hydroborate selectively the more reactive double bond of a diene, as in the present preparation, or double bonds in the presence of reducible groups, such as ethyl 10-undecenoate (**P8.5**). Terminal double bonds react readily at room temperature, so that the higher temperature utilized for the trisubstituted olefin in **P4.4** is not required. The lower temperature also favors the selective reaction desired.

3. For many purposes the organoborane may be used without isolation; however, it can be isolated by removal of the solvent and vacuum distillation under nitrogen **(S9.7)**, bp 103 at 0.035 mm, 21.3 g (93%).
4. The product can be oxidized **(P4.4:N3)** to the alcohol **(P4.2)**. It can be utilized to introduce the 2-(4-cyclohexenyl)ethyl group into the α-position of esters, ketones, nitriles, and related derivatives **(S7.9)**. Its conversion into the corresponding aldehyde is readily achieved[1] **(S7.5)**.

CH_2CH_2-B ⟶ CH_2CH_2CHO

[1] H. C. Brown, E. F. Knights, and R. A. Coleman, *J. Amer. Chem. Soc.*, **91**, 2144 (1969).

4.6. B-CYCLOPROPYL-9-BORABICYCLO[3.3.1]NONANE

$$H-C\equiv C-CH_2Br + 2\ B-H \xrightarrow[\text{8 hr}]{\text{pentane, 25°}} H-C(B)(B)-CH_2-CH_2Br$$

$$H-C(B)(B)-CH_2-CH_2Br \xrightarrow[-78°]{MeLi/Et_2O} B-\text{cyclopropyl} + B-Me + LiBr$$

Procedure by Stanley P. Rhodes[1] and Charles G. Scouten[2]

Procedure

An oven-dried 300 ml flask equipped with a septum inlet, pressure-equalized addition funnel, magnetic stirring bar, and mercury bubbler is flushed with nitrogen, and a static pressure of the gas is maintained throughout the reaction **(S9.1)**. Solid 9-BBN **(P2.11** or Aldrich) is introduced into the flask, 24.4 g (0.20 mole) **(P4.4:N1)**. The system is again flushed with nitrogen, and 100 ml of dry, olefin-free pentane is added. The flask is immersed in a cold-water bath, and 11.9 g (0.10 mole) propargyl bromide **(N1)** is added slowly via syringe to the stirred slurry. After one hour, the cold-water bath is removed, and the reaction mixture is allowed to stir 8 hours at room temperature. The reaction mixture is cooled to −78°C via an external Dry Ice-acetone cold bath. Standardized methyllithium solution in ether, 50 ml of 2.0*M* (0.10 mole) **(S9.10)** is added to the funnel using a graduated cylinder and a double-ended needle **(S9.2, F9.17) (N8)** and then dripped into the reaction mixture. The mixture is stirred about 15 minutes at −78° after all of the methyllithium has been added, then allowed to warm to room temperature. After stirring 1 hour, 25 ml of water is added. The organic layer is decanted using a double-ended needle into an evacuated distillation assembly **(F9.31) (N3)** where the pentane is flash distilled off **(S9.7)**. The residual oil is distilled at ∿ 15 mm giving 9.5 g (70%) of B-methyl-9-BBN (bp 64-65° at 15 mm) and 11.3 g (70%) of B-cyclopropyl-9-BBN (bp 101-104° at 15 mm) **(N3, N4, N5, N6)**.

Notes

1. Commercial, bp 82° (743 mm) n^{20}_{D} 1.4906.
2. Alternatively, the methyllithium solution can be added slowly through the septum inlet using a small gauge, double-ended needle. If this is done, the addition funnel may be omitted.
3. An efficient distillation column, such as a micro Widmer or spinning band, is essential in separating the two organoboranes.
4. The ^{1}H NMR exhibited cyclopropyl protons: 1H at 0.60 ppm: 4H at 0.55 ppm multiplets; ^{11}B NMR: −84 ppm (THF).
5. Oxidation of a sample **(S9.10)** yielded cyclopropanol, bp 100-103°, and *cis*-1,5-cyclooctandiol, mp 72-73°.
6. In the original publication of this preparation,[1] aqueous sodium hydroxide was used to effect the closure. The present procedure gives higher yields and affords easier purification. Methyllithium is essential for the related closures:[1,2]

B
|
$$H{-}C{-}(CH_2)_nOTs \xrightarrow{MeLi} B(CH_2)_n$$
|
B

$n = 4, 5, 6$

[1] H. C. Brown and S. P. Rhodes, *J. Amer. Chem. Soc.*, 91, 4306 (1969).
[2] C. G. Scouten, Ph.D. thesis, Purdue University, 1974.

4.7. B-*p*-TOLYL-9-BORABICYCLO[3.3.1] NONANE

$$\text{9-BBN (BH)} + p\text{-}CH_3C_6H_4Li \xrightarrow[0^\circ]{THF} [\text{9-BBN}(H)(p\text{-}C_6H_4CH_3)]^- Li^+$$

$$[\text{9-BBN}(H)(p\text{-}C_6H_4CH_3)]^- Li^+ + CH_3SO_3H \xrightarrow[0^\circ]{THF} \text{9-BBN}\text{–}C_6H_4\text{–}CH_3 + H_2\uparrow + CH_3SO_3Li\downarrow$$

Procedure by Milorad M. Rogić[1]

Procedure

Two representative sets of apparatus are assembled and flushed with nitrogen **(P2.5)**. In both cases the flasks are fitted with a septum inlet. One apparatus is used for the preparation of *p*-tolyllithium in ether solution in the usual manner from excess lithium, 3.06 g (0.440 mole), and 34.2 g (24.6 ml, 0.200 mole) of *p*-bromotoluene[2] **(N1)**. In the second apparatus is placed 200 ml of 0.50*M* 9-BBN in tetrahydrofuran **(P2.11** or Aldrich) and the flask and its contents cooled to 0°. *p*-Tolyllithium (0.10 mole) in ether is taken up in a hypodermic syringe and added to the solution of 9-BBN **(N2)**. Immediately following the addition, 9.6 g, (6.5 ml, 0.100 mole) of methanesulfonic acid is added. Hydrogen is rapidly evolved **(N3)**. The salt (lithium methanesulfonate) is allowed to settle, and the clear solution transferred by double-ended needle under nitrogen to a dropping funnel attached to a small distilling flask **(S9.7)**. The solvents are removed at atmospheric pressure, and the residue (containing lithium bromide) concentrates in the distilling flask, and is distilled under vacuum **(S9.7, P2.7)**. There is obtained 18.0 g (85% yield) of B-*p*-tolyl-9-borabicyclo-[3.3.1]nonane as a colorless liquid, bp 155-160° at 5.5 mm **(N4, N5)**.

Notes

1. The resulting organolithium reagent is standardized prior to use **(S9.10)**.
2. Originally, it was assumed that the reaction proceeds simply to the formation

of the trisubstituted borohydride. In most instances this does not appear to be the case. Consequently, the above mechanism must be considered tentative at this time.

3. A total of 2.5 ℓ (0.100 mole) of hydrogen is evolved and should be safely vented. The reaction of methanesulfonic acid with the parent compound, 9-BBN, is relatively slow under these conditions, requiring almost 1 hour for complete hydrogen evolution. On the other hand, the intermediate reacts rapidly, hydrogen being evolved almost instantly as acid is added.
4. A high pot temperature (∿220°C) is necessary to drive the product away from the solid salts. A heat gun or free-flame may be used.
5. The same procedure was used to prepare B-phenyl-9-BBN, B-methyl-9-BBN, and B-*n*-butyl-9-BBN. An alternative procedure involving the reaction of B-methoxy-9-BBN or, more generally, R_2BOCH_3, with alkylithiums has been developed[3] (**S3.2, P4.8**).

[1] H. C. Brown and M. M. Rogić, *J. Amer. Chem. Soc.*, **91**, 4304 (1969).
[2] H. Gilman, E. A. Zoellner, and W. M. Selbey, *J. Amer. Chem. Soc.*, **54**, 1957 (1932).
[3] G. W. Kramer and H. C. Brown, *J. Organomet. Chem.*, 73, 1 (1974).

4.8. B-*TERT*-BUTYL-3,5-DIMETHYLBORINANE

$$\text{3,5-dimethylborinane (B–H)} + CH_3OH \xrightarrow[0^\circ]{THF} \text{B-OCH}_3\text{-3,5-dimethylborinane} + H_2$$

$$\text{B-OCH}_3\text{-3,5-dimethylborinane} + LiC(CH_3)_3 \xrightarrow[-78^\circ \text{ to } 25^\circ]{\text{pentane}} \text{B-C(CH}_3)_3\text{-3,5-dimethylborinane} + LiOCH_3 \downarrow$$

Procedure by Ei-ichi Negishi[1] and Gary W. Kramer[2]

Procedure

3,5-Dimethylborinane, 0.200 mole, in tetrahydrofuran is prepared as described in **P2.12**. To the solution is added 9.6 g (0.300 mole) of methanol to form the methoxy derivative (**N1**). Pressure is reduced, and the tetrahydrofuran and excess methanol are removed. The flask and its contents are returned to atmospheric pressure with nitrogen (**N2**). Pentane, 200 ml, is added and the solution cooled to −78°. Then 100 ml of 2.0*M* *tert*-butyllithium is added by the double-ended needle technique to the well-stirred mixture (**S9.2**). Excess must be avoided (**S9.10**). At this point, the solution is clear. After 30 minutes, the bath is removed, and the temperature is allowed to rise to 25°. A precipitate, lithium methoxide, slowly forms and is allowed to settle overnight without stirring. The supernatant liquid is transferred under nitrogen by double-ended needle into an evacuated distilling flask (**F9.33**). The solid is washed twice with pentane, and the washings and reaction product slowly run into the distilling flask with concurrent distillation of the pentane. Residual solvent is removed under vacuum, and the product is distilled (**S9.7**). There is obtained 29.2 g (88% yield) of B-(*tert*-butyl)-3,5-dimethylborinane, bp 72-73° (11 mm) (**N3, N4**).

Notes

1. Hydrogen is rapidly evolved (5.0 ℓ, 0.200 mole). It should be directed away and safely vented.
2. If desired, the B-methoxy-3,5-dimethylborinane can be recovered by distillation under vacuum (nitrogen bleed): bp 59-60° (18 mm).
3. After oxidation of a sample (**S9.10; P2.7:N7**), VPC examination revealed the presence of 98% *tert*-butyl alcohol.
4. B-*tert*-butyl-3,5-dimethylborinane and similar derivatives have been used to achieve 1,4-additions to α-β-unsaturated carbonyl derivatives[1] (**P6.13, P8.24**).

$$\text{B-}(CH_3)_3C\text{-3,5-dimethylborinane} + CH_2{=}CHCOCH_3 \xrightarrow[25^\circ]{H_2O} (CH_3)_3CCH_2CH_2COCH_3$$

[1] E. Negishi and H. C. Brown, *J. Amer. Chem. Soc.*, **95**, 6757 (1973).
[2] G. W. Kramer and H. C. Brown, *J. Organomet. Chem.*, 73, 1 (1974).

4.9. (+)- and (–)-2-BUTANOL

$$(\text{Ipc})_2BH + CH_3CH{=}CHCH_3 \xrightarrow{\text{DG, } 0^\circ} (\text{Ipc})_2B\text{–}\overset{*}{C}H(CH_3)CH_2CH_3$$

$$(\text{Ipc})_2B\text{–}\overset{*}{C}H(CH_3)CH_2CH_3 + NaOH + 3\,H_2O_2 \xrightarrow{40^\circ} HO\text{–}\overset{*}{C}H(CH_3)CH_2CH_3 + 2\,\text{IpcOH} + NaB(OH)_4$$

Procedure by George Zweifel and Nagaraj R. Ayyangar[1]

Procedure

(+)-Diisopinocampheylborane (0.100 mole) in diglyme is prepared **(P2.1)**. The flask and its contents under static nitrogen pressure is immersed in an ice-salt bath and cooled to approximately −10°. *cis*-2-Butene is added in slight excess (6.2 g. 0.110 mole) **(N1)**, and the reaction mixture is stirred for 4 hours at 0°. Water, 10 ml, is added to destroy residual hydride **(N2)**. The temperature is raised to 25°, 33 ml of 3*M* sodium hydroxide is rapidly added and the oxidation carried out by the slow addition of 33 ml of 30% hydrogen peroxide **(P2.1:N6)** maintaining the temperature below 40° **(N3)**. After stirring for an additional hour, the alcohols formed are extracted with 200 ml of ether. The ether extract is washed once with saturated sodium chloride, and then dried over anhydrous magnesium sulfate. The ether is removed and the alcohol distilled through an efficient micro column. There is obtained 6.2 g of (R)-2-butanol (84% yield), bp 98° (752 mm), n^{20}D 1.3970, $[\alpha]^{26}$D +11.7°, indicating an optical purity of 87% **(N4)**.

In a related preparation (−)-diisocampheylborane from (+)-α-pinene **(N5)** reacts with *cis*-2-butene to yield 90% of (S)-2-butanol, bp 98° (744 mm), n^{20}D 1.3975, $[\alpha]^{20}$D −11.8°, an optical purity of 87% **(N4)**.

Notes

1. The procedure described in **P2.7** can be used to introduce the 2-butene. Alternatively, 9.9 ml (6.2 g) of the *cis*-2-butene can be condensed in a graduated glass tube at −78° and the material volatilized from the tube into the diglyme solution of the reagent **(S9.3)**.
2. If the hydroboration has been successful, very little hydrogen will be evolved. By monitoring the hydrogen evolved with a gas meter, it is possible to follow such reactions readily.
3. The static nitrogen pressure must be maintained until after completion of the oxidation **(P2.1:N7)**. Consequently, both the base and the 30% hydrogen peroxide should be introduced using a hypodermic syringe or double-ended needle, without opening the system.
4. The highest notation reported for (−)-2-butanol is [α]D −13.5°.[2]
5. (−)-α-pinene is available from the Aldrich Chemical Company. The (+)-α-pinene was a sample distilled from French turpentine and was originally supplied by the Hercules Powder Company. It is also presently available from Aldrich.

[1] H. C. Brown, N. R. Ayyangar, and G. Zweifel, *J. Amer. Chem. Soc.*, **86**, 397 (1964).
[2] P. J. Leroux and H. J. Lucas, *J. Amer. Chem. Soc.*, **73**, 41 (1951).

4.10. 1,3,2-BENZODIOXABOROLE(CATECHOLBORANE)

$$\text{C}_6\text{H}_4(\text{OH})_2 + \text{H}_3\text{B:THF} \xrightarrow[0^\circ]{\text{THF}} \text{C}_6\text{H}_4\text{O}_2\text{BH} + 2\,\text{H}_2$$

Procedure by Shyam K. Gupta[1]

Procedure

A 1-ℓ round-bottomed flask containing a magnetic stirring bar is fitted with a connecting tube attached to a mercury bubbler **(S9.1)**. The connecting tube should contain a sidearm fitted with a rubber stopple so situated that liquids can be added to or removed from the flask with a hypodermic syringe or a double-ended needle **(S9.2)**. The apparatus is flushed with nitrogen as the parts are flamed with a Bunsen burner **(N1)**. The nitrogen is conveniently introduced through a hypodermic needle inserted through the rubber stopple, and a static nitrogen pressure is maintained throughout the preparation **(S9.1)**. The flask is immersed in an ice bath, and 550 ml of 2.0*M* borane-tetrahydrofuran complex **(P2.3)** (1.10 mole, 10% excess) is introduced through the sidearm inlet with the aid of a double-ended needle **(P2.4:N1)**. Pure, thoroughly dry catechol **(N2)**, 110 g, 1.00 mole, is dissolved in 200 ml of dry tetrahydrofuran **(S9.11)**, and the solution is slowly added to the borane over 4 to 6 hours **(N3)**. After the addition is complete, the solution is stirred until no more hydrogen is evolved. A dropping funnel connected to a Claisen head, attached to a 2 ft Vigreaux column and a 300-ml flask, is connected to a fraction collector **(P4.7)** and flushed with nitrogen. A double-ended needle is used to transfer the solution to the dropping funnel. The solution is then run into the distilling flask as the tetrahydrofuran is removed under aspirator pressure (40 to 50 mm) at room temperature **(N4)**. After all of the tetrahydrofuran has been removed, the catecholborane is recovered by careful distillation. There is obtained 96 g (79% yield) of catecholborane, bp 76-77° (100 mm) $n^{20}\text{D}$ 1.5070 **(N5)**.

Notes

1. Alternatively, the parts can be dried in an oven at 125° and the apparatus assembled hot **(S9.1)**.
2. High purity catechol, dried over phosphorous pentaoxide in a dessicator, mp 105°.
3. Hydrogen is evolved vigorously as the catechol solution reacts with the borane. The reaction liberates 50 ℓ (2.00 mole) of hydrogen. This should be conducted away and safely vented. If a gas meter is available, it is desirable

to measure the hydrogen as an additional indication that the reaciton is proceeding satisfactorily.

4. It is desirable to remove almost all of the tetrahydrofuran at the reduced pressure and room temperature. Otherwise, an appreciable fraction of tetrahydrofuran and catecholborane will co-distill initially.
5. Catecholborane is a liquid at room temperature and solidifies in an ice bath. It is quite stable at room temperature, but appears to deteriorate slowly over several weeks. It keeps far better as the solid in a cold room, and this is the recommended procedure for storing the reagent over relatively long periods of time. The neat material is 9.0*M* in catecholborane.

The product can be standardized in the usual manner by hydrolysis (**S9.10**).

[1]H. C. Brown and S. K. Gupta, *J. Amer. Chem. Soc.*, 93, 1816 (1971).

4.11. *TRANS*-CYCLOHEXYLETHENYLBORONIC ACID

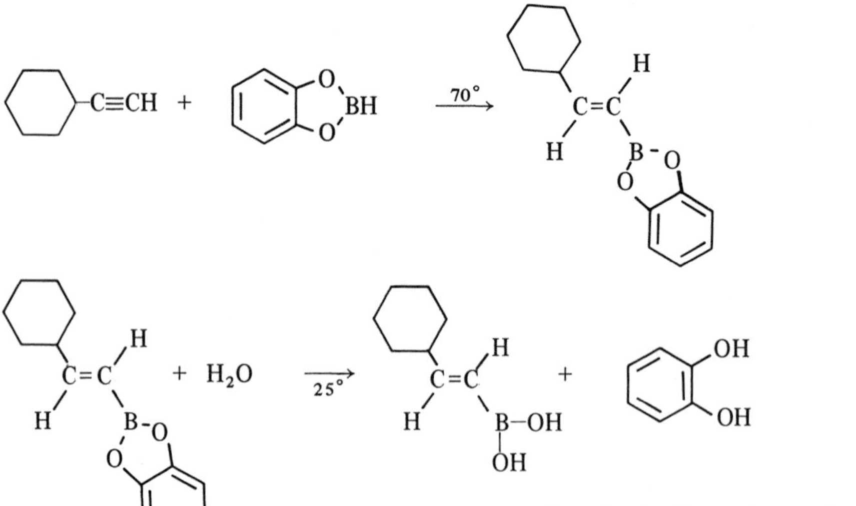

Procedure by Shyam K. Gupta[1]

Procedure

A 200-ml flask containing a magnetic stirring bar is connected to a mercury bubbler by means of a tube containing a sidearm fitted with a rubber stopple (**S9.1**). A stream of nitrogen is introduced through a hypodermic needle penetrating the rubber stopple, and the apparatus is dried by flaming (**P4.10:N1**). The apparatus is allowed to cool, and a static nitrogen pressure is maintained.

With the aid of a hypodermic syringe, 10.8 g (0.100 mole) of cyclohexylethyne (**N1**) is introduced, followed by 12.2 g (0.100 mole) of catecholborane (**P4.10**). The stirred mixture is heated for 1 hour at 70° (**N2**). The product is cooled to room temperature, and 100 ml of water is added. The white crystalline product formed is filtered and recrystallized from water to give 15.0 g (97% yield) of *trans*-cyclohexylethenylboronic acid, mp 104-105° (**N3, N4**).

Notes

1. Commercial, bp 132° (740 mm), n^{20}D 1.4597. It may be purified by a distillation from a small amount of sodium borohydride (**S9.11**).
2. The product can be distilled: 2-(*trans*-cyclohexylethenyl)-1,3,2-benzodioxaborole, bp 114° (0.2 mm), n^{20}D 1.5430, a yield of 93% (VPC), 82% isolated.
3. These vinylboronic acids are stable to air and can be handled by the usual methods.
4. The reaction can be extended to internal acetylenes. These are less reactive and require approximately 4 hours at 70°. The reaction exhibits a high degree of regioselectivity, with the boron atom adding to the less hindered position of the triple bond. For example, 1-cyclohexylpropyne reacts to give a 92:8 distribution of isomers:

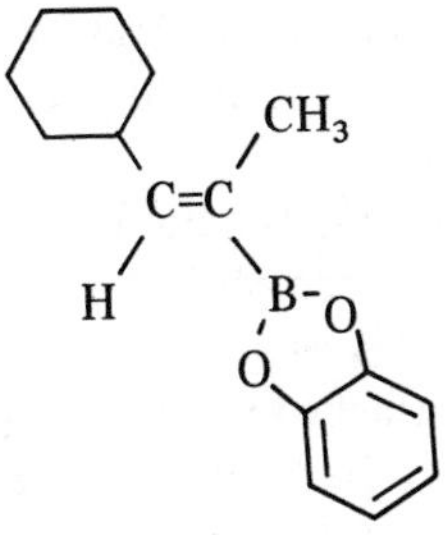

CH$_3$ C=C H B O O

92%

[1] H. C. Brown and S. K. Gupta, *J. Amer. Chem. Soc.*, **94**, 4370 (1972).

4.12. DICHLOROBORANE AND MONOCHLOROBORANE DIETHYL ETHERATES

$$LiBH_4 + 3\,BCl_3 + 4\,Et_2O \xrightarrow[0^\circ]{EE} 4\,Cl_2BH{:}OEt_2 + LiCl\downarrow$$

$$LiBH_4 + BCl_3 + 2\,Et_2O \xrightarrow[0^\circ]{EE} 2\,ClBH_2{:}OEt_2 + LiCl\downarrow$$

Procedure by Nair Ravindran[1,2]

Procedure

1.00*M* solutions of lithium borohydride and boron trichloride in ethyl ether are prepared and mixed to produce the desired chloroborane. The lithium chloride precipitates and can be removed by filtration. For some applications, the presence of the lithium chloride offers no difficulty, and it need not be removed.

The lithium borohydride solution is prepared as follows: A 200-ml, two-neck round-bottomed flask containing a magnetic stirring bar is fitted to a connecting tube leading to a mercury bubbler **(N1)**. The second neck is closed by a tube fitted with a rubber stopple to permit introduction and removal of liquids with a hypodermic syringe. Lithium borohydride **(N2)**, 2.50 g (0.110 mole allowing for 95% purity) is placed in the flask, and the system is flushed with nitrogen. Then a static nitrogen pressure is maintained. The flask is cooled with an ice-water bath, and 110 ml of anhydrous ethyl ether **(N3)** is added to the flask with the aid of a hypodermic syringe while the contents of the flask are stirred. The cold bath is removed and the stirring continued for 2 hours. A small amount of white material fails to dissolve **(N4, N5)**.

The solution of boron trichloride in ethyl ether is prepared as follows: A dry apparatus with a 500-ml flask is assembled in a well-ventilated fume hood and flushed with nitrogen. Then a static pressure of nitrogen is maintained. The flask is cooled in an ice-water bath, and either 100 ml (for monochloroborane) or 300 ml (for dichloroborane) of ether is introduced with the aid of a hypodermic syringe.

A quantitative transfer of boron trichloride into the ethyl ether without exposure to the atmosphere is accomplished in the following manner **(S9.3)**: A graduated tube is closed with a pressure tight stopper carrying an inlet and outlet tube. The outlet tube is attached to a mercury bubbler through a 15-gauge hypodermic needle. The inlet tube is attached to one arm of a T-tube. The second arm of the T-tube is connected to a nitrogen line. The third arm of the T-tube is connected to the exit valve of the boron trichloride cylinder **(N6)**. The system is flushed with dry nitrogen. Then the graduated tube is immersed in a -20° bath as the valve of the cylinder is gradually opened. There is collected 8.7 ml (0.100 mole) for monochloroborane or 26.1 ml (0.300 mole) for dichloroborane.

The hypodermic needle on the outlet tube is lifted from the mercury bubbler and quickly inserted into the flask through the rubber stopple such that the tip of the needle is about 1 cm above the surface of the ether. The -20° bath is now removed, and the boron trichloride is allowed to distill slowly into the flask while the ether within is stirred at 0°. When all of the boron trichloride has transferred, the hypodermic needle is withdrawn, and the flask is shaken in order to dissolve small quantities of solid $Cl_3B{:}OEt_2$ clinging to the sides of the flask above the liquid level **(N7, N8)**.

To prepare the monochloroborane etherate, a hypodermic syringe is used to withdraw 100 ml of the 1.00*M* solution of lithium borohydride in ethyl ether. This solution is added slowly to 100 ml of the 1.00*M* solution of boron trichloride in ethyl ether stirred at 0°. As the lithium borohydride is added, lithium chloride precipitates from the solution. After the addition is complete, the mixture is stirred for 2 hours at 0°. For many hydroborations this solution with precipitated lithium chloride can be used directly **(P4.14)**. On standing overnight, the lithium chloride settles, leaving a clear supernatant solution (1.0*M*) of $ClBH_2{:}OEt_2$ in ethyl ether **(N9)**. If desired, this can be filtered, as described for dichloroborane. In contrast to dichloroborane etherate, monochloroborane etherate solutions cannot be concentrated to obtain the neat product.

For the preparation of dichloroborane etherate the procedure is the same, except that 300 ml of 1.00*M* boron trichloride is prepared. The 100-ml of 1.00*M* lithium borohydride in the syringe is added slowly to this solution stirred at 0°. After the addition is complete, the mixture is stirred for 2 hours at 0°. It is allowed to stand overnight at 0° to allow the lithium chloride to settle, leaving a supernatant solution of $Cl_2BH{:}OEt_2$ **(N9)**. [To ensure the absence of minor amounts of monochloroborane, it is desirable at this point to add 5% excess (0.015 mole) of boron trichloride etherate. Alternatively, this excess can be present at the time the lithium borohydride solution is added.]

A receiver flask, 500-ml, to be used for reaction or storage, is surmounted by a 500-ml medium porosity filter chamber, and the system is flushed with nitrogen **(S9.6, F9.27)**. The dichloroborane solution is transferred under nitrogen pressure through an 18-gauge double-ended needle to the filter funnel and then into the flask. Two successive portions of ethyl ether are injected into the original flask to wash the residual lithium chloride and then transferred through the filter into the receiver. The filter chamber is removed under a steady flow of nitrogen to protect the product, and the flask is immediately closed with a connector tube leading to a water aspirator (protected so that water vapor cannot reach the product). The ether is removed at 10 mm and 25° until the flask and its contents have reached constant weight. In this way an essentially quantitative yield of dichloroborane diethyl etherate, 62.8 g (0.40 mole), is obtained (containing about 5% of excess $Cl_3B{:}OEt_2$ to repress formation of monochloroborane). The product is a colorless liquid, mp −25 to −30°. In the presence of boron trichloride it converts olefins into the corresponding $RBCl_2$ derivatives[3] **(P4.13)** **(N10)**. It is miscible with benzene, carbon tetrachloride, and ethyl ether, but not with pentane or other paraffinic solvents.

Notes

1. All glassware is dried in an oven at 125° and assembled hot, or flame-dried under a stream of dry nitrogen **(S9.1)**.

2. Lithium borohydride, 95%, from Alfa Chemical Company. Lithium borohydride is a light powder which has tendency to fly around. Cellulosic materials, such as tissue paper, can ignite on contact with lithium borohydride in air. Lithium borohydride should be handled with the same precautions used for lithium aluminum hydride. Lithium borohydride will dissolve in ether to give 1.1 to 1.2*M* solutions **(S9.11)**.
3. We found it satisfactory to use Mallinckrodt anhydrous ethyl ether stored over Type 4A molecular sieves under nitrogen **(S9.11)**.
4. The insoluble material can be removed by filtration under nitrogen. However, this is not essential for these preparations.
5. The solution can be standardized by injecting an aliquot of the pure supernatant liquid into a hydrolyzing mixture of 1:1 glycerol and water at 0° and measuring the hydrogen evolved **(S9.10)**.
6. Boron trichloride, >99%, is available from Matheson Gas Products.
7. The solubility of boron trichloride etherate in ethyl ether is approximately 1.5*M* at 0°. The boron trichloride solution in ether may be stored in a cold room for several weeks without observable change. The concentration of such solutions can be conveniently determined by hydrolyzing an aliquot and titrating the hydrochloric acid produced with standard base, using methyl orange as indicator **(S9.10)**.
8. If several preparations utilizing the chloroboranes are involved, it is convenient to prepare stock solutions of lithium borohydride and boron trichloride in ether and store them in a cold room.
9. The solution should be analyzed for hydride **(N5)**, chloride **(N7)**, and boron **(S9.10)**. If the hydride and chloride are not in the proper ratio of 2:1 or 1:2 a calculated quantity of the solution in deficiency is added.
10. At 25° a white suspension forms within 2 days. However, in the cold room (0°), the material has been kept 2 weeks as a clear liquid which can be utilized successfully.

[1] H. C. Brown and N. Ravindran, *J. Amer. Chem. Soc.*, **94**, 2112 (1972).
[2] H. C. Brown and N. Ravindran, *J. Amer. Chem. Soc.*, **95**, 2396 (1973).
[3] H. C. Brown and P. A. Tierney, *J. Inorg. Nucl. Chem.*, 9, 51 (1959).

4.13. CYCLOPENTYLDICHLOROBORANE; DIMETHYL CYCLOPENTYLBORONATE

$$\text{cyclopentene} + Cl_2BH{:}OEt_2 + BCl_3 \xrightarrow[0^\circ]{\text{pentane}} \text{C}_5\text{H}_9\text{-}BCl_2 + Cl_3B{:}OEt_2\downarrow$$

$$\text{C}_5\text{H}_9\text{-}BCl_2 + 2\,CH_3OH \xrightarrow[0^\circ]{} \text{C}_5\text{H}_9\text{-}B(OCH_3)_2 + 2\,HCl$$

Procedure by Nair Ravindran[1]

Procedure

A dry **(P4.12:N1)** 300-ml round-bottomed flask fitted with a septum capped inlet, a magnetic stirring bar, and a connecting tube leading to a mercury bubbler is flushed with nitrogen and then maintained under a static pressure **(S9.1)**. The flask is immersed in an ice-water bath and charged with 125 ml of dry, olefin-free pentane, 6.8 g, (0.100 mole) of cyclopentene **(S9.11)**, and 50 ml of a 2.0*M* solution of boron trichloride (0.100 mole) in pentane **(N1)**. The mixture is stirred in the ice bath for 10 minutes and then 0.100 mole of $Cl_2BH{:}OEt_2$ **(P4.12)** is slowly added with a syringe over a period of 5 minutes. The reaction mixture is stirred at 0° for 15 minutes and then stirred for an additional 15 minutes at 25°. The flask is again cooled to 0°, and the pentane solution is transferred under a positive pressure of nitrogen into another flask (equipped in the same manner) through a glass tube fitted with a fritted disk **(F9.26)**. The solid $Cl_3B{:}OEt_2$ in the original reaction flask is washed with pentane (2 × 25 ml) and the washings transferred to the main solution. The pentane is then removed under vacuum by attaching a protected water aspirator to the outer end of the connecting tube closing the flask.

When all of the pentane has been removed, the connecting tube is replaced by a distillation assembly **(S9.7)**, and the pure cyclopentyldichloroborane is isolated by distillation **(N2)** under nitrogen, bp 136-138° (751 mm), 11.9 g (79% yield). The reaction appears to be broadly applicable to the preparation of $RCBl_2$ derivatives **(N3)**.

To prepare the ester, the reaction mixture above is not filtered. Instead, after stirring for 15 minutes at 25°, the reaction mixture is cooled to 0°, and 50 ml of methanol is injected slowly, while stirring vigorously **(N4)**. The stirring is continued for 30 minutes at 0° after completing the addition of methanol. The solvent, excess methanol, methyl borate (bp 68°), and residual hydrogen chloride are removed by evacuation with a protected water aspirator and the dimethyl cyclopentylboronate isolated by vacuum distillation, bp 76-78° (40 mm), 10.8 g (76% yield).

Notes

1. The reagents are transferred to the reaction flask through the septum-capped inlet using hypodermic syringes lubricated with Halocarbon oil **(S9.2)**.
2. Considerable foaming may occur during distillation. It is recommended that a foam trap be incorporated in the distillation assembly.
3. The R_2BCl and $RBCl_2$ derivatives are much more electrophilic than the corresponding R_3B compounds and greatly facilitate certain reactions **(P6.15, P8.16)**.
4. Hydrogen chloride is produced during the methanolysis and should be absorbed.

[1]H. C. Brown and N. Ravindran, *J. Amer. Chem. Soc.*, **95**, 2396 (1973).

4.14. DICYCLOPENTYLCHLOROBORANE; METHYL DICYCLOPENTYLBORINATE

$$2\ \text{C}_5\text{H}_8 + ClBH_2{:}OEt_2 \xrightarrow[0^\circ]{EE} (\text{C}_5\text{H}_9)_2BCl + Et_2O$$

$$(\text{C}_5\text{H}_9)_2BCl + CH_3OH \xrightarrow[0^\circ]{CH_3OH} (\text{C}_5\text{H}_9)_2BOCH_3 + HCl$$

Procedure by Nair Ravindran[1]

Procedure

Monochloroborane-diethyl ether solution is prepared as described (**P4.12**). The entire solution can be utilized for hydroboration-oxidation (**N1**) without separation of the lithium chloride precipitate. The following procedure utilizes the supernatant solution above the precipitated salt. A dry 100-ml round-bottomed flask containing a magnetic stirring bar, equipped with a septum-capped inlet protected by a Teflon stopcock, and a connecting tube leading to a mercury bubbler is dried and flushed with nitrogen in the usual manner (**P4.12:N1**). The flask is immersed in an ice bath. Then 50.0 ml of the clear supernatant 1.00*M* $ClBH_2{:}OEt_2$ solution (**P4.12**) is withdrawn with a syringe (**N2**) and introduced into the above flask. Cyclopentene, 6.8 g (0.100 mole) **S9.11**, is added slowly with a syringe through the inlet tube to the stirred reaction mixture. The stirring is continued for one hour following the addition. The outer end of the connecting tube is then attached to a protected water aspirator and the ether removed at room temperature under aspirator pressure. The connecting tube is replaced by a vacuum distillation assembly (**N3**). Pure dicyclopentylchloroborane is readily isolated by distillation under reduced pressure, 7.4 g (80% yield), bp 68-69° (1.0 mm). This synthesis appears to be broadly applicable[1] (**P4.13:N3**).

The dialkylchloroboranes are readily converted into the borinate esters. After stirring for 1 hour following the addition of the cyclopentene, 3.2 g (0.100 mole, 100% excess) of methanol is injected into the reaction vessel, and the mixture is stirred for 15 minutes at 0°. The connecting tube is attached to a protected water aspirator, and the ether and dissolved hydrogen chloride are removed at room temperature. The connecting tube is replaced by a vacuum distillation assembly, and the pure methyl dicyclopentylborinate is isolated by distillation, 7.6 g (84% yield), bp 82-84° (2.0 mm) (**N4**).

Notes

1. Hydroboration with chloroborane is considerably more regiospecific than hydroboration with diborane[2] (**S3.5**). For oxidation of the products with

alkaline hydrogen peroxide, ethanol should be utilized as a cosolvent, and two equivalents of aqueous base are required.
2. Halocarbon oil is used to lubricate the syringe.
3. The product must be protected by a stream of nitrogen during the change. Both the dialkylchloroboranes and the corresponding methyl esters are sensitive to oxygen and water.
4. A major new application for the borinic acid esters is their ready transformation into the corresponding ketones (**S7.8**).

[1] H. C. Brown and N. Ravindran, *J. Amer. Chem. Soc.*, **94**, 2112 (1972).
[2] H. C. Brown and N. Ravindran, *J. Org. Chem.*, 38, 182 (1973).

4.15. BIS(*CIS*-3-HEXENYL)CHLOROBORANE

$$2\,H_5C_2C{\equiv}CC_2H_5 + ClBH_2{:}OEt_2 \xrightarrow[0^\circ]{EE} (cis\text{-}H_5C_2CH{=}CC_2H_5)_2BCl$$

H₅C₂ C₂H₅ / C=C / H BCl / H / C=C / H₅C₂ C₂H₅

Procedure by Nair Ravindran[1]

Procedure

In the flask of **P4.14** is placed 8.2 g (0.100 mole) of 3-hexyne in 15 ml of anhydrous ethyl ether (**N1**). Chloroborane in ether (**P4.12**), 50 ml of a 1.00*M* solution (0.05 mole), is slowly added by syringe to the stirred solution at 0°. The mixture is stirred under nitrogen for 2 hours at 0°. Then the ether is removed using a protected water aspirator, and the bis-(*cis*-3-hexenyl)chloroborane is recovered by distillation, 9.3 g (88% yield), bp 66-68° at 0.1 mm (**N2**).

Notes

1. In the case of terminal acetylenes, such as 1-hexyne, the use of a moderate excess (40%) of the acetylene minimizes dihydroboration and improves the yield.
2. The product can be protonolyzed with acetic acid to *cis*-3-hexene in 92% yield and oxidized with alkaline hydrogen peroxide to 3-hexanone in 96% yield. The most interesting application of these materials is their direct utilization in the Zweifel diene synthesis[2] (**P8.19**).

(Et)(H)C=C(Et)–BCl–C(Et)=C(H)(Et) $\xrightarrow[I_2]{NaOH}$ (Et)(H)C=C(Et)–C(Et)=C(H)(Et)

[1]H. C. Brown and N. Ravindran, *J. Org. Chem.*, **38**, 1617 (1973).
[2]G. Zweifel, N. L. Polston, and C. C. Whitney, *J. Amer. Chem. Soc.*, **90**, 6243 (1968).

4.16. B-THEXYL-4,8-DIMETHYL-2-BORABICYCLO[3.3.1] NONANE (THEXYLLIMONYLBORANE)

BH_2 + $\xrightarrow[0°]{THF}$ B

Procedure by Carl D. Pfaffenberger[1]

Procedure

A 200-ml, 3-neck flask fitted with two 50-ml dropping funnels and a magnetic stirring bar is connected to a mercury bubbler and dried. The apparatus is flushed with nitrogen and maintained under a static nitrogen pressure **(S9.1)**. In the flask is placed 20 ml of tetrahydrofuran. In one dropping funnel is placed 50 ml of a 1.00*M* solution of thexylborane in tetrahydrofuran **(P2.10)**. In the other dropping funnel is placed 50 ml of a 1.00*M* solution of limonene **(N1)**. The flask is immersed in a 0° bath, and the thexylborane and limonene are added slowly to the stirred flask over 1 hour. The reaction mixture is stirred for an additional hour and then brought to room temperature. The dropping funnels are removed in a stream of nitrogen and replaced by glass stoppers. A distilling assembly is attached **(N2)**. The tetrahydrofuran is removed at reduced pressure and room temperature. The pressure is further reduced, and the product, thexyllimonylborane **(N3)**, is collected without exposure to atmospheric oxygen, 8.4 g (72% yield), bp 88° (0.4 mm), $[\alpha]^{22}D$ + 168°.

Notes

1. Commercial D-(+)-limonene was distilled from a small quantity of lithium aluminum hydride (*Caution:* Foaming!), bp 55° (5 mm), $n^{20}D$ 1.4724, $[\alpha]^{20}D$ +118°.

2. The procedure can be simplified by preparing the required amount of thexylborane (**P2.10**), cooling the flask to −30°, and adding the equivalent of limonene by direct, rapid injection. The solution is brought to 0° and allowed to stir for 1 hour, then distilled as in the above procedure.
3. Many dienes can be successfully directed to achieve cyclic hydroboration with thexylborane (**S3.6**). These cyclic derivatives have interesting synthetic possibilities. For example, protonolysis of thexyllimonylborane, followed by oxidation, provides a stereospecific synthesis of (1R, 2R, 4S)-carvomenthol.[1] The derivative has been converted to 4,8-dimethylbicyclo[3.3.1]-nonan-2-one.[2] Finally, it has been converted to a borohydride which exhibits a remarkable stereoselectivity in the reduction of certain prostaglandin intermediates.[3]

[1] H. C. Brown and C. D. Pfaffenberger, *J. Amer. Chem. Soc.*, **89**, 5475 (1967).
[2] A. Pelter, M. G. Hutchings, and K. Smith, *Chem. Comm.*, 1048 (1971).
[3] E. J. Corey and R. K. Varma, *J. Amer. Chem. Soc.*, **93**, 7319 (1971).

4.17. THEXYL(*TRANS*-2-METHYLCYCLOPENTYL)BORANE; *TRANS*-2-METHYLCYCLOPENTYLBORANE-TRIETHYLAMINE

$$\text{ThexylBH}_2 + \text{1-methylcyclopentene} \xrightarrow[-25^\circ,\ 1\ \text{hr}]{\text{THF}} \text{thexyl}(trans\text{-2-methylcyclopentyl})\text{BH}$$

$$\text{thexyl}(trans\text{-2-methylcyclopentyl})\text{BH} + Et_3N \xrightarrow[1\ \text{hr}]{25^\circ} trans\text{-2-methylcyclopentyl-}BH_2{:}NEt_3 + (CH_3)_2C{=}C(CH_3)_2$$

Procedure by Ei-ichi Negishi and Jean-Jacques Katz[1]

Procedure

Thexylborane in tetrahydrofuran, 100 ml of 1.00*M*, is prepared in a 300-ml flask as described in **P2.10**. The solution (under nitrogen) is cooled to −25°. To the stirred solution is added 8.2 g (0.100 mole) of 1-methylcyclopentene (**P4.4: N2**). The solution is maintained at −25° for 1 hour to provide the thexyl(*trans*-2-methylcyclopentylborane in essentially quantitative yield (**N1**).

Such thexylmonoalkylboranes can serve to hydroborate a wide variety of olefins and acetylenes, including those containing functional groups (**P4.18, P8.4, P8.13**). These fully substituted thexyl derivatives have many valuable applications. However, of special interest and significance is the utilization of the thexylmonoalkylboranes to prepare the corresponding monoalkylborane-triethylaminates.

To the above solution of thexyl(*trans*-2-methylcyclopentyl)borane is added 40.4 g (0.400 mole) of triethylamine (excess), and the mixture is allowed to come to 25° and stirr at that temperature for 1 hour. Evaporation of the volatile components at 15 mm and room temperature (2 hours) produces *trans*-2-methylcyclopentylborane-triethylamine in nearly quantitative yield. This provides the first simple synthesis of monoalkylboranes **(N2)**.

Notes

1. Simple terminal olefins, such as 1-hexene, yield mixtures of mono- and dialkylated products. However, other types, such as 2-methyl-1-propene, cyclopentene, norbornene, and 2-methyl-2-butene, all react cleanly to give the thexylmonoalkylboranes. Even the highly hindered olefin, 2,3-dimethyl-2-butene, can be forced to react by the use of a high concentration, large excess, and long reaction time to yield dithexylborane, the first monomeric dialkylborane.[2]
2. Even though the monoalkylboranes are isolated as the triethylaminates, they are reactive. Evidently these addition compounds are partially dissociated at 25°. Thus treatment of the product with methanol at 25° yields the corresponding dimethyl *trans*-2-methylcyclopentylborinate. Treatment of siamylboranetriethylamine with two molar equivalents of cyclopentene yields siamyldicyclopentylborane.

[1] H. C. Brown, E. Negishi, and J.-J. Katz, *J. Amer. Chem. Soc.*, **94**, 5893 (1972).
[2] E. Negishi, J.-J. Katz, and H. C. Brown, *J. Amer. Chem. Soc.*, **94**, 4025 (1972).

4.18. THEXYLCYCLOHEXYLCYCLOPENTYLBORANE

BH_2 + [cyclohexene] $\xrightarrow[-25°, 1 hr]{THF}$ thexyl(cyclohexyl)borane (B–H)

thexyl(cyclohexyl)borane + [cyclopentene] $\xrightarrow[-25°, 24 hr]{(CH_3)_2C{=}C(CH_3)_2}$ thexylcyclohexylcyclopentylborane

Procedure by Jean-Jacques Katz[1]

Procedure

Thexylborane in tetrahydrofuran, 100 ml of 1.00*M*, is prepared as described in **P2.10**. The solution (under nitrogen) is cooled to −25°, and 8.2 g (0.100 mole)

of cyclohexene (**S9.11**) is added to prepare thexylmonocyclohexylborane (**P4.17**) (**N1**). After 1 hour, 8.4 g (0.100 mole) of 2,3-dimethyl-2-butene is added (**N2**), followed by 6.8 g (0.100 mole) of cyclopentene (**S9.11**). The reaction mixture is maintained at −25° for 24 hours (**N3, N4**).

Notes

1. Such thexylmonoalkylboranes readily hydroborate olefins and acetylenes without special precautions, even those with functional substituents (**P8.4, P8.18**).

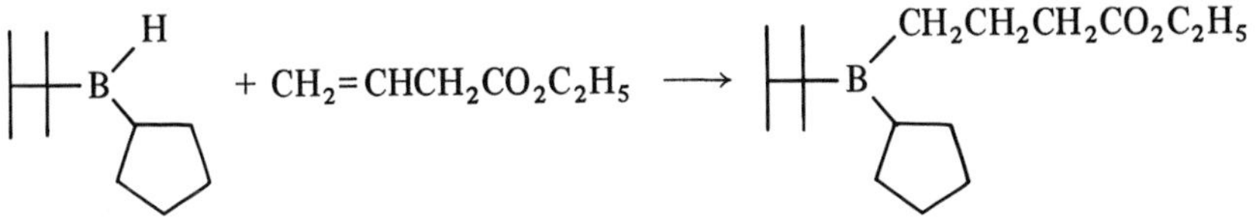

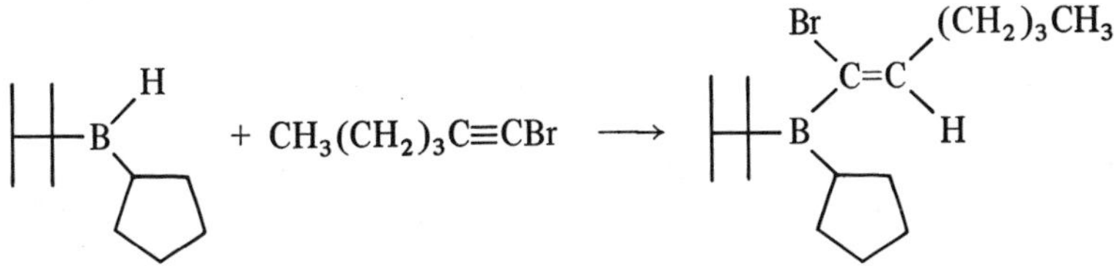

$$\text{Thexyl(cyclopentyl)BH} + CH_2{=}CHCH_2CO_2C_2H_5 \longrightarrow \text{Thexyl(cyclopentyl)B}{-}CH_2CH_2CH_2CO_2C_2H_5$$

$$\text{Thexyl(cyclopentyl)BH} + CH_3(CH_2)_3C{\equiv}CBr \longrightarrow \text{Thexyl(cyclopentyl)B}{-}C(Br){=}CH(CH_2)_3CH_3$$

   However, attempts to hydroborate olefins of greater steric requirements results in displacement of 2,3-dimethyl-2-butene.[2]

2. The difficulty in achieving a dihydroboration of thexylborane with olefins of relatively large steric requirements apparently arises from the existence of an equilibrium.[1]

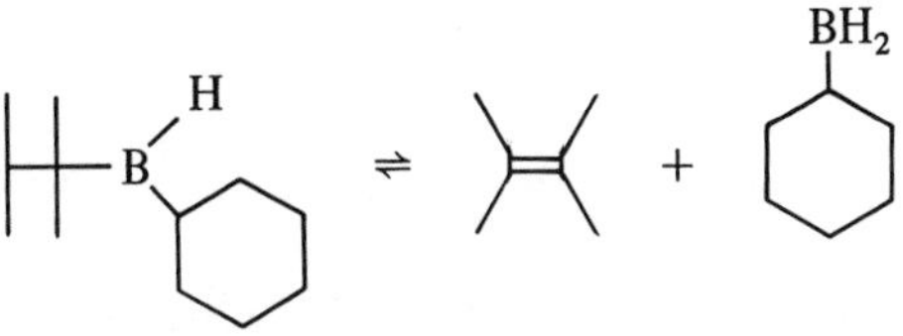

$$\text{Thexyl(cyclohexyl)BH} \rightleftharpoons \text{2,3-dimethyl-2-butene} + \text{cyclohexyl-}BH_2$$

   This equilibrium can be utilized to produce the monoalkylboranes[3] (**P4.17**). It can be repressed by adding additional 2,3-dimethyl-2-butene (present procedure) to permit the complete hydroboration.[1]

3. In less hindered systems, a reaction time of 4 hours is adequate.
4. Oxidation of the product (**P6.10**) produces approximately 0.100 mole each of thexyl alcohol, cyclohexanol, and cyclopentanol. The homogenity of the structure of such trialkylated boranes has been demonstrated by conversion to the trialkylcarbinylboronates[3] (**P8.9**) and the corresponding trialkylcarbinols, RR′R″COH[4] (**P6.10**).

[1]H. C. Brown, J.-J. Katz, and E. Negishi, manuscript in preparation.
[2]C. F. Lane and H. C. Brown, *J. Organomet. Chem.*, **34**, C29 (1972).
[3]H. C. Brown, E. Negishi, and J.-J. Katz, *J. Amer. Chem. Soc.*, **94**, 5893 (1972).
[4]H. C. Brown, B. A. Carlson, and J.-J. Katz, *J. Org. Chem.*, **38**, 3968 (1973).

5

ORGANOBORANE CONVERSIONS: SURVEY

The hydroboration reaction[1] has made the organoboranes readily available. Investigation has revealed that these are truly remarkable reagents, possibly the most versatile available to the organic chemist.[2-4] The fact that it is now possible to synthesize organoboranes with many functional groups (**S1.7**) and to replace the boron-carbon bond by many substituents (**S5**) or by carbon-carbon bonds (**S7**) opens up a major new approach in synthetic chemistry.

The present section discusses those reactions of organoboranes that do not involve the formation of carbon-carbon bonds. Reactions leading to the formation of such bonds are discussed subsequently (**S7**).

5.1. ISOMERIZATION OF ORGANOBORANES

At moderate temperatures the organoboranes undergo a facile isomerization that proceeds to place the boron atom predominantly at the least hindered position of the alkyl groups.[5]

[1] H. C. Brown, *Hydroboration*, W. A. Benjamin, New York, 1962.
[2] H. C. Brown, *Boranes in Organic Chemistry*, Cornell University Press, Ithaca, N.Y., 1972.
[3] G. M. L. Cragg, *Organoboranes in Organic Synthesis*, Marcel Dekker, New York, 1973.
[4] C. F. Lane, *Aldrichimica Acta*, 6, 21 (1973).
[5] H. C. Brown and G. Zweifel, *J. Amer. Chem. Soc.*, 88, 1433 (1966).

This facile migration of the boron atom makes possible a number of valuable syntheses that are not otherwise practical. For example, oxidation of the initial hydroboration product from 3-hexene yields pure 3-hexanol, whereas oxidation of the thermally treated product yields predominantly 1-hexanol. Similarly the following transformations are readily achieved in essentially quantitative yield **(P6.1)**.

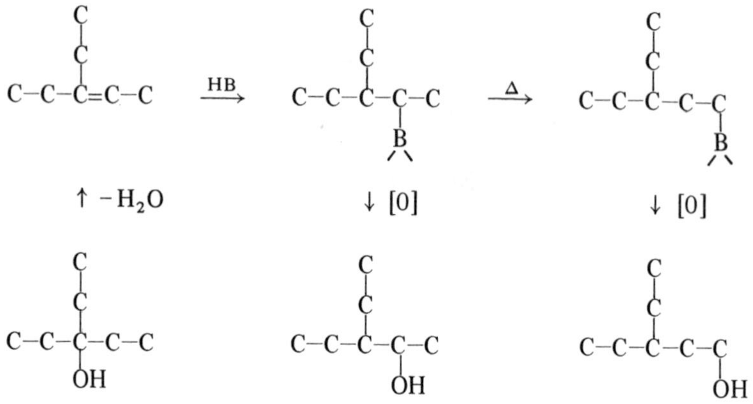

Hydroboration of *β*-pinene yields tri(*cis*-myrtanyl)borane. Thermal treatment of the product converts it into the more stable *trans*-myrtanyl derivative.[6] It is also possible to move the boron atom from the ring to the side chain.[7]

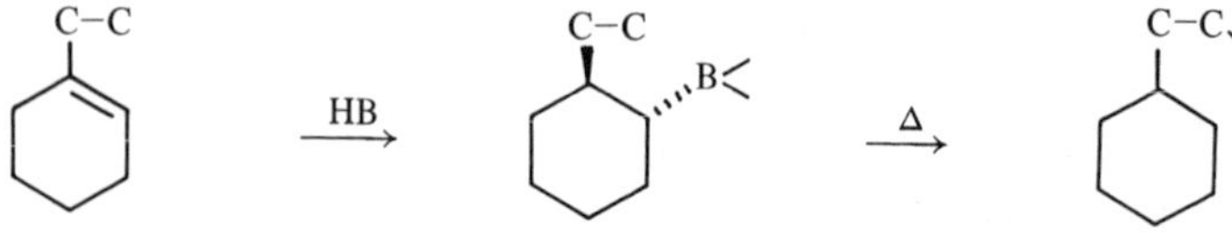

Such ready isomerization makes possible a number of interesting syntheses of desired compounds from alternative, more readily available intermediates, such as the synthesis of the exocyclic alcohol, 2-cyclohexylethanol, from the readily available endocyclic olefin, 1-ethylcyclohexene.

5.2. DISPLACEMENT REACTIONS OF ORGANOBORANES

The mechanism of the isomerization reaction appears to involve a partial dissociation of the organoborane into olefin and a boron-hydrogen moiety, followed by readdition. The process occurs repeatedly, until the boron atom

[6] G. Zweifel and H. C. Brown, *J. Amer. Chem. Soc.*, 86, 393 (1964).
[7] H. C. Brown and G. Zweifel, *J. Amer. Chem. Soc.*, 89, 561 (1967).

ends up at the least hindered position of the molecule, thereby yielding the most stable of the organoborane derivable from the particular alkyl groups used.[8]

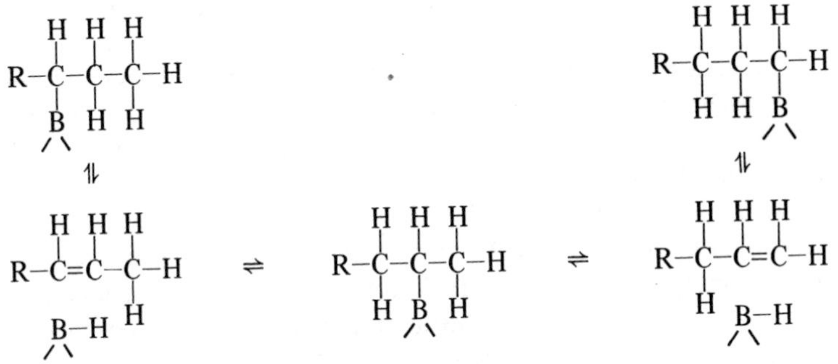

```
    H H H                                    H H H
    | | |                                    | | |
  R-C-C-C-H                                R-C-C-C-H
    | | |                                    | | |
    B H H                                    H H B
   / \                                          / \
    ⇅                                          ⇅
    H H H              H H H                 H H H
    | | |              | | |                 | | |
  R-C=C-C-H    ⇌     R-C-C-C-H     ⇌       R-C-C=C-H
        |              | | |                 |
   B-H  H              H B H                 H  B-H
  / \                   / \                    / \
```

It is evident from this mechanism that the presence of another olefin, of equal or greater reactivity, should result in the displacement of the original olefin from the organoborane.[8,9]

5.3. CONTRATHERMODYNAMIC ISOMERIZATION OF OLEFINS

The combination of hydroboration, isomerization, and displacement provides a practical synthetic route for the contrathermodynamic isomerization of olefins[10] **(P6.2)**.

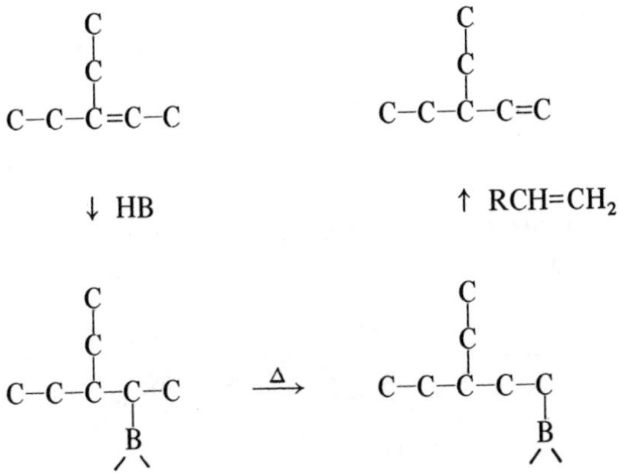

```
        C                          C
        |                          |
        C                          C
        |                          |
  C-C-C=C-C                  C-C-C-C=C

      ↓ HB                   ↑ RCH=CH2

        C                          C
        |                          |
        C                          C
        |                          |
  C-C-C-C-C        Δ→        C-C-C-C-C
          |                          |
          B                          B
         / \                        / \
```

[8]H. C. Brown and B. C. Subba Rao, *J. Org. Chem.*, **22**, 1136 (1957).
[9]R. Köster, *Ann.*, **618**, 31 (1958).
[10]H. C. Brown and M. V. Bhatt, *J. Amer. Chem. Soc.*, **88**, 1440 (1966).

The process can also be utilized to move the double bond from the endocyclic to the exocyclic position.[11]

5.4. CYCLIZATION

On thermal treatment trialkylboranes, such as tri-*n*-pentylborane, liberate olefin and hydrogen with the formation of cyclic species.[12] The reaction apparently involves a thermal dissociation into the dialkylborane, followed by reaction of >B–H with H–C≤ .

Now that hydroboration with thexylborane can be controlled to give the thexylmonoalkylborane (**S3.6, P4.17**), it is possible to proceed directly to the dialkylborane intermediate. If the alkyl group is structurally capable of participating in the formation of a ring structure, simple thermal treatment of the adduct gives such cyclic boron derivatives[13] (**P6.3**).

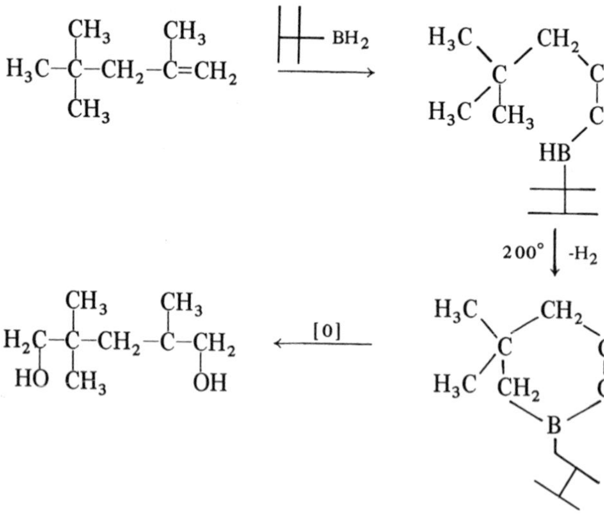

$$H_3C-C(CH_3)_2-CH_2-C(CH_3)=CH_2 \xrightarrow{\text{ThexylBH}_2} H_3C-C(CH_3)_2-CH_2-CH(CH_3)-CH_2-BH\text{-Thexyl} \xrightarrow[-H_2]{200^\circ} \text{cyclic borane (B-thexyl)} \xrightarrow{[O]} HOCH_2-C(CH_3)_2-CH_2-CH(CH_3)-CH_2OH$$

It is possible to achieve more difficult cyclizations under more drastic thermal conditions (250-350°).[14]

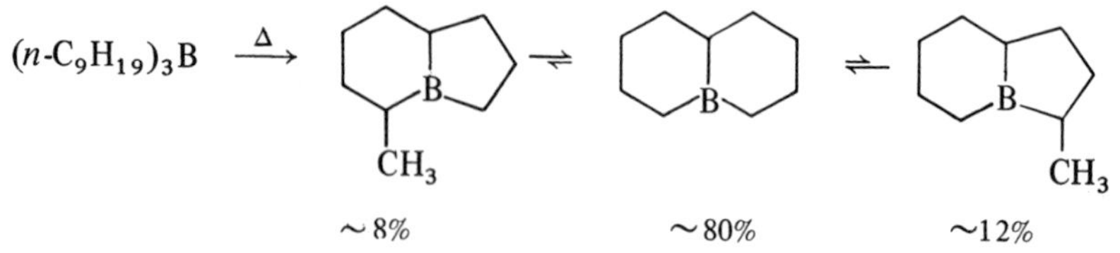

$$(n\text{-}C_9H_{19})_3B \xrightarrow{\Delta} \text{methyl-substituted bicyclic borane } (\sim 8\%) \rightleftharpoons \text{boradecalin } (\sim 80\%) \rightleftharpoons \text{methyl-substituted bicyclic borane } (\sim 12\%)$$

[11] H. C. Brown, M. V. Bhatt, T. Munekata, and G. Zweifel, *J. Amer. Chem. Soc.*, **89**, 567 (1967).
[12] P. F. Winternitz and A. A. Carothi, *J. Amer. Chem. Soc.*, **82**, 2430 (1960).
[13] H. C. Brown, K. J. Murray, H. Muller, and G. Zweifel, *J. Amer. Chem. Soc.*, **88**, 1443 (1966).
[14] R.Köster, *Angew. Chem. Internat. Edit.*, **3**, 174 (1964).

Hydroboration provides an alternative, milder procedure to such derivatives.[15]

$$\begin{array}{l} CH_2CH{=}CH_2 \\ | \\ CH_2CH{=}CH_2 \end{array} + H_2B{-}\text{thexyl} \longrightarrow \begin{array}{l} CH_2CH_2CH_2 \\ | \qquad\qquad\quad B{-}\text{thexyl} \\ CH_2CH_2CH_2 \end{array}$$

$$CH_2{=}CHCH_2CH{=}CHCH_2CH_2CH{=}CH_2 \xrightarrow[\Delta]{BH_3} \text{(bicyclic borane, B)}$$

5.5. PROTONOLYSIS

The organoboranes are remarkably stable to water, aqueous bases, and aqueous mineral acids. However, they are susceptible to protonolysis by carboxylic acids. Trialkylboranes require relatively vigorous conditions. In general, they must be heated under reflux with excess propionic acid in diglyme for 2 to 3 hours.[16] Vinyl boranes are much more reactive, often undergoing rapid protonolysis with acetic acid at 0°[17] **(P6.5)**.

Thus one can take advantage of the unique properties of the hydroboration reaction[6] **(S1.4)** to achieve stereospecific hydrogenations **(P6.4)**.

CH_2 (β-pinene) $\xrightarrow{HB}$ $CH_2{-})_3B$ (cis) $\xrightarrow[DG]{EtCO_2H}$ CH_3 *cis*-Pinane

$CH_2{-})_3B$ (cis) $\xrightarrow{\Delta}$ $CH_2{-})_3B$ (trans) $\xrightarrow[DG]{EtCO_2H}$ CH_3 *trans*-Pinane

The reaction can be utilized to achieve hydrogenations which would be difficult with noble metal catalysts.

[15] H. C. Brown and E. Negishi, *J. Amer. Chem. Soc.*, **91**, 1224 (1969).
[16] H. C. Brown and K. Murray, *J. Amer. Chem. Soc.*, **81**, 4108 (1959).
[17] H. C. Brown and G. Zweifel, *J. Amer. Chem. Soc.*, **83**, 3834 (1961).

$$RSCH_2CH{=}CH_2 \xrightarrow{HB} RSCH_2CH_2CH_2{-}B\lt \xrightarrow[DG]{EtCO_2H} RSCH_2CH_2CH_3$$

Protonolysis appears to proceed with retention of configuration. This makes possible the synthesis of deuterium derivatives with known stereochemistry.[17,18]

RCO$_2$D

RC≡CR → (R)(H)C=C(R)(B–) $\xrightarrow{RCO_2D}$ (R)(H)C=C(R)(D)

The reaction can be utilized to synthesize *cis*-enynes and *cis,cis*-dienes.[19]

R–C≡C–C≡C–R $\xrightarrow{R'_2BH}$ (R)(H)C=C(C≡CR)(BR'$_2$) $\xrightarrow{CH_3CO_2H}$ (R)(H)C=C(C≡CR)(H)

↓

(R)(H)C=C(BR'$_2$)–C(H)=C(BR'$_2$)(R) $\xrightarrow{CH_3CO_2H}$ (R)(H)C=C(H)–C(H)=C(H)(R)

5.6. HALOGENOLYSIS

The carbon-boron bond in trialkylboranes is surprisingly stable to the direct action of halogens, such as bromine and iodine.[2] In the presence of alkali, reaction readily occurs to give the corresponding alkyl halides.[20,21]

For example, iodine fails to react directly with organoboranes except under relatively drastic conditions.[22] However, the addition of sodium hydroxide in methanol to the organoborane and iodine brings about a rapid reaction.[20]

[18] H. C. Brown and K. J. Murray, *J. Org. Chem.*, **26**, 631 (1961).
[19] G. Zweifel and N. L. Polston, *J. Amer. Chem. Soc.*, **92**, 4068 (1970).
[20] H. C. Brown, M. W. Rathke, and M. M. Rogić, *J. Amer. Chem. Soc.*, **90**, 5038 (1968).
[21] H. C. Brown and C. F. Lane, *J. Amer. Chem. Soc.*, **92**, 6660 (1970).
[22] L. H. Long and D. Dollimore, *J. Chem. Soc.*, 3902, 3906 (1953).

$$(RCH_2CH_2)_3B + 2\,I_2 + 2\,NaOH \longrightarrow 2\,RCH_2CH_2I + 2\,NaI + RCH_2CH_2B(OH)_2$$

Only two of three primary alkyl groups react, and the situation is even less favorable for secondary alkyl groups. Fortunately, the use of disiamylborane circumvents this difficulty and provides excellent yields for terminal olefins.

$$RCH{=}CH_2 + Sia_2BH \longrightarrow RCH_2CH_2BSia_2$$

$$RCH_2CH_2BSia_2 + I_2 + NaOH \longrightarrow RCH_2CH_2I + Sia_2BOH + NaI$$

The reaction **(P6.6)** is widely applicable, as indicated by the following representative transformations.

H₃C–C(=CH₂)–C₆H₅ ⟶ H₃C–CH(CH₂I)–C₆H₅

(β-pinene, =CH₂) ⟶ (H, CH₂I)

CH=CH₂ (cyclohexenyl) ⟶ CH₂CH₂I (cyclohexenyl)

The reaction of bromine with trialkylboranes is also greatly facilitated by alkali. However, the presence of water must be avoided. Apparently, in the presence of aqueous alkali, hypobromite is formed, and this converts organoboranes into alcohols. The use of sodium methoxide in methanol solves the problem.[21] In the case of primary alkyl groups, all three react **(P6.7)**.

$$(CH_3CH_2CH_2\overset{CH_3}{\overset{|}{C}}HCH_2)_3B + 3\,Br_2 + 4\,NaOCH_3 \longrightarrow 3\,CH_3CH_2CH_2\overset{CH_3}{\overset{|}{C}}HCH_2Br + 3\,NaBr + NaB(OH)_4$$

99%

$$CH_3O_2C(CH_2)_8CH{=}CH_2 \longrightarrow CH_3O_2C(CH_2)_8CH_2CH_2Br$$

92%

Unexpectedly, in the case of the *exo*-norbornyl derivative, the reaction proceeds with inversion of configuration to form *endo*-norbornyl bromide.[23]

$\xrightarrow{Br_2, NaOCH_3}$

Br

Since such *endo*-derivatives are exceedingly difficult to synthesize, this route may prove valuable.

Finally, secondary alkyl bromides are readily synthesized by the action of bromine on the B-alkyl-9-BBN derivatives.[24]

B $\xrightarrow[0° \text{ to } 25°, 1.5 \text{ hr}]{Br_2, CH_2Cl_2}$ Br 84%

B $\xrightarrow[0° \text{ to } 25°, 1.5 \text{ hr}]{Br_2, CH_2Cl_2}$ Br 90%

Interestingly, this reaction does not proceed through a direct rupture of the boron-carbon bond, but involves an initial substitution of the α-hydrogen by bromine (**S7.10**).

Vinylic boron derivatives can be converted into the corresponding halides.[25,26] Thus the treatment of a vinylboronic acid (from the hydrolysis of the corresponding catechol derivative) with iodine and base results in the replacement of the boronic acid by iodine with retention of configuration[26] (**P6.8**).

[23] H. C. Brown and C. F. Lane, *Chem. Commun.*, 522 (1971).
[24] C. F. Lane and H. C. Brown, *J. Organomet. Chem.*, **26**, C51 (1971).
[25] A. F. Kluge, K. G. Untch, and J. H. Fried, *J. Amer. Chem. Soc.*, **94**, 7827 (1972).
[26] H. C. Brown, T. Hamaoka, and N. Ravindran, *J. Amer. Chem. Soc.*, **95**, 5786 (1973).

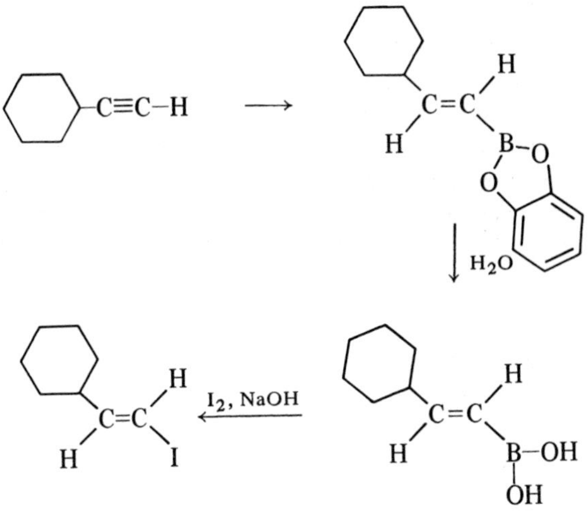

On the other hand, the addition of bromine followed by alkali, yields the *cis*-vinyl bromide, with inversion of configuration[27] **(P6.9)**.

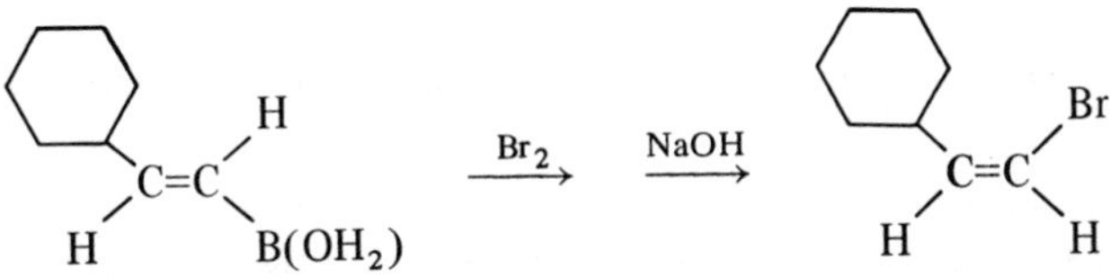

The precise mechanisms have not yet been established. However, it is evident that these are valuable reactions for the synthesis of vinyl halides of known configurations.

5.7. OXIDATION-ALKALINE HYDROGEN PEROXIDE

The reaction of alkaline hydrogen peroxide with organoboranes[28] is a remarkably clean reaction of wide generality.

$$R_3B + 3\,H_2O_2 + NaOH \longrightarrow 3\,ROH + NaB(OH)_4$$

[27]H. C. Brown, T. Hamaoka, and N. Ravindran, *J. Amer. Chem. Soc.*, **95**, 6456 (1973).
[28]J. R. Johnson and M. G. Van Campen, Jr., *J. Amer. Chem. Soc.*, **60**, 121 (1938).

It is essentially quantitative and proceeds with clean retention of configuration. It takes place readily in the presence of the usual hydroboration media, such as diglyme, tetrahydrofuran, and ethyl ether. [For water-insoluble solvents, such as ethyl ether, and for intermediates not easily oxidized (**P6.10**), it is desirable to add ethanol as a cosolvent.] It can accommodate all groups which tolerate hydroboration. Consequently, hydroboration followed by *in situ* oxidation with alkaline hydrogen peroxide provides a simple, broadly applicable procedure to achieve the anti-Markovnikov hydration of double bonds.[29]

CH_3 $\xrightarrow{HB}$ $\xrightarrow{[O]}$ OH H H CH_3

$\xrightarrow{HB}$ $\xrightarrow{[O]}$ OH

$$RO_2CCH_2CH{=}CH_2 \xrightarrow{HB} \xrightarrow{[O]} RO_2CCH_2CH_2CH_2OH$$

Vinylic boranes also undergo the reaction, providing a means of converting triple bonds into the corresponding carbonyl derivatives[17,30] (**P4.3**).

$$CH_3(CH_2)_5C{\equiv}CH \xrightarrow{Sia_2BH} CH_3(CH_2)_5(H)C{=}C(H)BSia_2 \xrightarrow{[O]} CH_3(CH_2)_6CHO$$

$C{\equiv}C{-}CH_3$ $\longrightarrow$ $C{=}C$ CH_3 H B–O O $\xrightarrow{[O]}$ CH_2COCH_3

[29] G. Zweifel and H. C. Brown, *Org. React.*, 13, 1 (1964).
[30] H. C. Brown and S. K. Gupta, *J. Amer. Chem. Soc.*, 94, 4370 (1972).
[31] H. C. Brown and C. P. Garg, *J. Amer. Chem. Soc.*, 83, 2951 (1961).

5.8. OXIDATION-CHROMIC ACID

Organoboranes can be oxidized directly to ketones with aqueous chromic acid.[31] Yields in the range of 65-85% are realized (**P6.11**).

This reaction provides a convenient general route for the following transformation.

Hydroborations of certain internal olefins yield two isomeric boron derivatives, which are oxidized to two isomeric ketones.[32]

The less hindered of these two ketones is readily separated with sodium bisulfite, providing a simple route to such ketones.[32]

5.9. OXIDATION-OXYGEN

The organoboranes react readily with oxygen to produce mixtures of hydroperoxides and alcohols.[2] Careful control of the oxidation conditions leads to the predominant formation of $RB(O_2R)_2$.[33]

$$R_3B + 2\,O_2 \longrightarrow RB(O_2R)_2$$

Treatment of the product with aqueous alkali results in a reaction of the re-

[32] H. C. Brown, I. Rothberg, and D. L. Vander Jagt, *J. Org. Chem.*, **37**, 4098 (1972).
[33] H. C. Brown, M. M. Midland, and G. W. Kabalka, *J. Amer. Chem. Soc.*, **93**, 1024 (1971).

maining carbon-boron bond with one of the alkyl hydroperoxides formed in the hydrolysis.

$$RB(O_2R)_2 + 3\,H_2O + NaOH \longrightarrow 2\,ROH + RO_2H + NaB(OH)_4$$

The introduction of the stoichiometric amount of oxygen, readily achieved with the automatic gasimeter **(S9.4)**, provides essentially quantitative conversion to alcohols.

$$R_3B + 1.5\,O_2 \longrightarrow R_{1.5}B(O_2R)_{1.5} \xrightarrow[H_2O]{NaOH} 3\,ROH$$

This procedure provides an alternative route for the conversion of organoboranes to alcohols. However, the reaction does not possess the stereospecificity of the alkaline peroxide procedure. For example, whereas the organoborane from norbornene is converted by alkaline hydrogen peroxide to 99.6% exo-norborneol, the oxygen procedure yields 86% *exo*-, 14% *endo*-. The loss of stereospecificity is doubtless a result of the free-radical nature of the reaction with oxygen.[34]

$$R_3B + O_2 \longrightarrow R\cdot$$

$$R\cdot + O_2 \longrightarrow RO_2\cdot$$

$$RO_2\cdot + R_3B \longrightarrow RO_2BR_2 + R\cdot, \text{ etc.}$$

Treatment of the oxidation intermediate with hydrogen peroxide in the absence of base oxidizes the remaining boron-carbon bond.

$$RB(O_2R)_2 + H_2O_2 + 2\,H_2O \longrightarrow ROH + 2\,RO_2H + B(OH)_3$$

Extraction with alkali separates the hydroperoxide from the alcohol, providing a convenient route to such alkyl hydroperoxides[35] **(P6.12)**.

$$\text{cyclohexene} \xrightarrow{HB} \xrightarrow{O_2} \xrightarrow{H_2O_2} \text{cyclohexyl}-O_2H$$

The yields of alkyl hydroperoxide are excellent, based on the conversion of two groups to the desired product. Thus the yield of cyclohexyl hydroperoxide is 95% on the basis of two groups, or 63% based on the cyclohexene used.

[34] A. G. Davies and B. P. Roberts, *J. Chem. Soc. B*, 311 (1969).
[35] H. C. Brown and M. M. Midland, *J. Amer. Chem. Soc.*, 93, 4078 (1971).

Fortunately, it is now possible to overcome this difficulty for cases where the olefin is a valuable intermediate which should be converted more completely to product. The reaction of oxygen with the ethyl etherates of the alkyldichloroboranes (**S3.5, P4.13**) proceeds to the complete conversion of the alkyl groups to the hydroperoxide.[26]

$$c\text{-}C_6H_{11}BCl_2{:}OEt_2 \xrightarrow{O_2} \xrightarrow{H_2O} c\text{-}C_6H_{11}O_2H$$

93%

5.10. SULFURIDATION

A detailed study of the conversion of the carbon-boron bond of organoboranes to the carbon-sulfur bond does not appear to have been made.[37] However, it has been observed that trialkylboranes will participate in facile photochemical chain reactions with organic disulfides, producing the corresponding thioethers[38] (**P6.13**).

$$R_3B + C_6H_5SSC_6H_5 \longrightarrow RSC_6H_5 + R_2BSC_6H_5$$

$$R_3B + 2\,CH_3SSCH_3 \longrightarrow 2\,RSCH_3 + RBS(CH_3)_2$$

The reaction appears to be broadly applicable, readily accommodating groups such as norbornyl which resist S_N2 substitution processes. The utilization of only one (in the phenyl disulfide reaction) or two alkyl groups (in the methyl disulfide reaction) can be circumvented by using the 3,5-dimethylborinane derivatives (**S3.2, P4.8**).

$$\text{B-cyclohexyl-3,5-dimethylborinane} + CH_3SSCH_3 \xrightarrow{h\nu} c\text{-}C_6H_{11}SCH_3$$

94%

[36]M. M. Midland and H. C. Brown, *J. Amer. Chem. Soc.*, **95**, 4069 (1973).
[37]Z. Yoshida, T. Okushi, and O. Manobe, *Tetrahedron Lett.*, 1641 (1970).
[38]H. C. Brown and M. M. Midland, *J. Amer. Chem. Soc.*, **93**, 3291 (1971).

5.11. AMINATION

Organoboranes react with chloramine to produce the corresponding primary amines.[39]

$$R_3B \xrightarrow{H_2NCl} RNH_2$$

Unfortunately, the reagent is unstable and must be freshly prepared before use. The yields in the preparation of the reagent are only moderate (about 50%) and somewhat erratic. Consequently, such solutions should be standardized before being used. These disadvantages suggested the desirability of a search for a more satisfactory reagent.

Fortunately, the relatively stable material, hydroxylamine-0-sulfonic acid, proved to be applicable to this synthesis. A highly satisfactory procedure has been developed to permit the simple synthesis of a wide variety of amines from the corresponding olefins via hydroboration[39,40] **(P6.14)**.

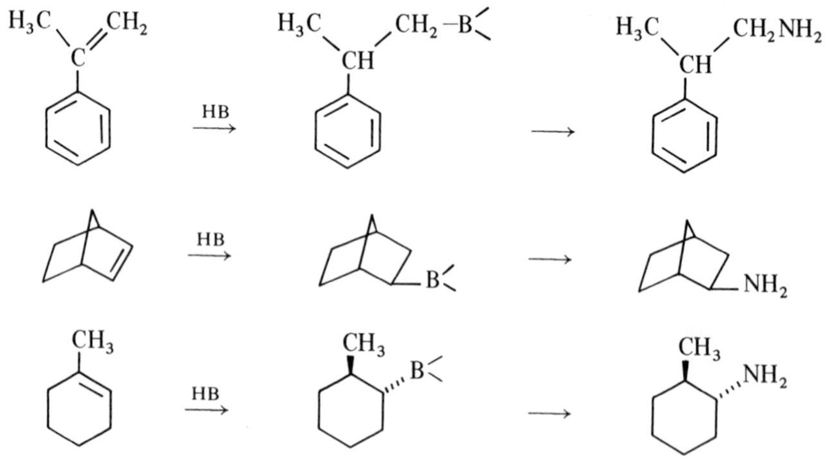

$$C_2H_5O_2C(CH_2)_8CH{=}CH_2 \xrightarrow{HB} \longrightarrow C_2H_5O_2C(CH_2)_8CH_2CH_2NH_2$$

The various organoborane derivatives, R_3B,[41] R_2BCl,[42] and $RBCl_2$,[43] all

[39] H. C. Brown, W. R. Heydkamp, E. Breuer, and W. S. Murphy, *J. Amer. Chem. Soc.*, **86**, 3565 (1964).

[40] M. W. Rathke, N. Inoue, K. R. Varma, and H. C. Brown, *J. Amer. Chem. Soc.*, **88**, 2870 (1966).

[41] A. Suzuki, S. Sono, M. Itoh, H. C. Brown, and M. M. Midland, *J. Amer. Chem. Soc.*, **93**, 4329 (1971).

[42] H. C. Brown and M. M. Midland, *J. Amer. Chem. Soc.*, **94**, 2114 (1972).

[43] H. C. Brown, M. M. Midland, and A. B. Levy, *J. Amer. Chem. Soc.*, **95**, 2394 (1973).

react with organic azides to give secondary amines. The reactions with the dialkylchloroboranes or the alkyldichloroboranes are especially facile.

$$R_2BCl + R'N_3 \xrightarrow{25\text{-}110^\circ} N_2 + RR'NBClR \xrightarrow{NaOH} RR'NH$$

$$RBCl_2 + R'N_3 \xrightarrow{25\text{-}80^\circ} N_2 + RR'NBCl_2 \xrightarrow{NaOH} RR'NH$$

The reaction proceeds with retention of configuration. Consequently, it is possible to utilize the stereospecificity of the hydroboration stage[43] **(P6.15)**.

HBCl2 → BCl2 → N3 → H N

The reaction makes possible a stereospecific synthesis of *N*-alkyl- and *N*-aryl-aziridines[44] **(P6.16)**.

H CH3 C=C H3C H IN3 → H N3 CH3 C–C CH3 I H PhBCl2 → -N2 RLi → Ph N H CH3 CH3 H

H H C=C H3C CH3 IN3 → H N3 H C–C H3C I CH3 PhBCl2 → -N2 RLi → Ph N H H H3C CH3

It would be desirable to have a similar general synthesis of tertiary amines. It was originally reported that tri-*n*-butylborane did not react with *N*-chloro-dimethylamine to give *n*-butyldimethylamine,[45] in a manner related to the corresponding reaction with chloroamine. Instead, the reaction proceeded to give *n*-butyl chloride in yields of 30 to 50%.

It now appears that this reaction is the result of a free radical chain.[46] Inhibition of the chain process by galvinoxyl allows the reaction to follow the polar path, producing *n*-butyldimethylamine.

$$n\text{-}Bu_3B + ClN(CH_3)_2 \xrightarrow{\text{galvinoxyl}} n\text{-}BuN(CH_3)_2 + n\text{-}Bu_2BCl$$

Unfortunately, the synthetic possibilities of this reaction have not yet been explored.

[44] A. B. Levy and H. C. Brown, *J. Amer. Chem. Soc.*, **95**, 4067 (1973).
[45] J. G. Sharefkin and H. D. Banks, *J. Org. Chem.*, **30**, 4313 (1965).
[46] A. G. Davies, S. C. W. Hook, and B. P. Roberts, *J. Organomet. Chem.*, **23**, C11 (1970).

5.12. METALATION

A major application of the Grignard reagent is in the synthesis of other organometallics, such as diethylmercury and tetraethyllead. The initial studies indicated that the organoboranes offer promise for such applications.[47]

The mercurials were selected for a detailed study of the utility of this approach to the synthesis of organometallics. Organoboranes derived from terminal olefins via hydroboration undergo a quantitative reaction in a matter of minutes with mercuric acetate at 0° or at room temperature to give the corresponding alkylmercuric acetates in essentially quantitative yields[48] **(P6.17)**.

$$(RCH_2CH_2)_3B + 3\,Hg(OAc)_2 \xrightarrow{THF} 3\,RCH_2CH_2HgOAc + B(OAc)_3$$

These alkylmercuric acetates are readily converted into the dialkylmercurials.[49]

$$2\,RCH_2CH_2HgOAc \xrightarrow[B(OAc)_3]{Zn} (RCH_2CH_2)_2Hg$$

Secondary alkyl groups require much more vigorous conditions to react. Consequently, it is possible to use the high selectivity of dicyclohexylborane to prepare the corresponding organomercurials.

CH=CH$_2$ (on cyclohexene ring) $\xrightarrow{(C_6H_{11})_2BH}$ CH$_2$CH$_2$BR$_2$ (on cyclohexene ring) $\xrightarrow{Hg(OAc)_2}$ CH$_2$CH$_2$HgOAc (on cyclohexene ring)

$$RO_2C(CH_2)_8CH{=}CH_2 \longrightarrow RO_2C(CH_2)_8CH_2CH_2HgOAc$$

The conversion of the vinyl derivatives of catecholborane proceeds stereospecifically and in essentially quantitative yield[50] **(P6.18)**.

R(H)C=C(R')–B(catecholate) $\xrightarrow{Hg(OAc)_2}$ R(H)C=C(R')HgOAc

[47] J. B. Honeycutt, Jr., and J. M. Riddle, *J. Amer. Chem. Soc.*, **82**, 3051 (1960).
[48] R. C. Larock and H. C. Brown, *J. Amer. Chem. Soc.*, **92**, 2467 (1970).
[49] J. D. Buhler and H. C. Brown, *J. Organomet. Chem.*, **40**, 265 (1972).
[50] R. C. Larock, S. K. Gupta, and H. C. Brown, *J. Amer. Chem. Soc.*, **94**, 4371 (1972).

The reaction has been used in a synthetic approach to the prostaglandins.[51]

5.13. SUMMARY

The discussion thus far has dealt with conversions that involved the formation of carbon-boron bonds or the transformation of such bonds into bonds with elements other than carbon. These transformations may be summarized schematically as indicated.

Isomerization	$-C_a-B<$	$\longrightarrow$	$-C_b-B<$
Displacement	$R-C-C-B<$	$\longrightarrow$	$R'-C-C-B<$
Cyclization	$-C-H\ H-B<$	$\longrightarrow$	$-C-B< + H_2$
Protonolysis	$-C-B<$	$\longrightarrow$	$-C-H + RCO_2B<$
Halogenolysis	$-C-B<$	$\longrightarrow$	$-C-X + CH_3O-B<$
Oxidation (H_2O_2 + NaOH)	$-C-B<$	$\longrightarrow$	$-C-OH + HO-B<$
Oxidation (H_2CrO_4)	$-C(H)-B<$	$\longrightarrow$	$>C{=}O + HO-B<$
Oxidation (O_2)	$-C-B<$	$\longrightarrow$	$-C-O_2H + HO-B<$
Sulfuridation	$-C-B<$	$\longrightarrow$	$-C-SR + RS-B<$
Amination (XNH_2)	$-C-B<$	$\longrightarrow$	$-C-NH_2 + HO-B<$
Amination (RN_3)	$-C-B<$	$\longrightarrow$	$-C-NHR + HO-B<$

[51] R. Pappo and P. W. Collins, *Tetrahedron Lett.*, 2627 (1972).

Metalation $\quad -\overset{|}{\underset{|}{C}}-B\lt \longrightarrow -\overset{|}{\underset{|}{C}}-MX + X-B\lt$

Transformations involving the conversion of boron-carbon bonds to carbon-carbon bonds are considered in **S7**.

6

ORGANOBORANES CONVERSIONS: PROCEDURES

6.1. 3-ETHYL-1-PENTANOL

$$8\,(C_2H_5)_2C{=}CHCH_3 + 3\,NaBH_4 + 4\,BF_3 \xrightarrow[25^\circ]{DG} 4\,[(C_2H_5)_2CHCH(CH_3)]_2BH + 3\,NaBF_4$$

$$[(C_2H_5)_2CHCH(CH_3)]_2BH + (C_2H_5)_2CH{=}CHCH_3 \xrightarrow[160^\circ,\ 2\ hr]{DG} [(C_2H_5)_2CHCH_2CH_2]_3B$$

$$[(C_2H_5)_2CHCH_2CH_2]_3B + 3\,H_2O_2 + NaOH \xrightarrow[40^\circ]{DG} 3\,(C_2H_5)_2CHCH_2CH_2OH + NaB(OH)_4$$

Procedure by George Zweifel[1]

Procedure

A 300-ml flask fitted with a septum inlet is equipped with a magnetic stirring bar, reflux condenser, and a mercury bubbler. The dry apparatus is flushed with nitrogen and then maintained under a static nitrogen atmosphere until after the oxidation (**S9.1**). Into the flask is injected with a syringe 14.7 g (0.150 mole) of 3-ethyl-2-pentene (**N1**), followed by 45 ml of a 1.00*M* solution of sodium borohydride in diglyme (20% excess) (**N2**). Hydroboration is achieved over 0.5

hour by the addition to the stirred solution at 25° of 16.4 ml of 3.65 *M* boron trifluoride (0.060 mole) in diglyme (**N3**). The solution is now heated for 2 hours at 160° (gentle reflux). The reaction mixture is brought to room temperature, and 10 ml of water is cautiously added to destroy residual hydride. When hydrogen is no longer evolved, 17 ml of 3 *M* sodium hydroxide is added rapidly. The oxidation is carried out by the slow addition of 17 ml of 30% hydrogen peroxide (exothermic reaction, **P2.1:N6**), maintaining the temperature below 40 to 50°. The reaction product is taken up in 100 ml of ether, and the ether extract is washed five times with equal volumes of ice water to remove the diglyme. The ether extract is dried over anhydrous magnesium sulfate. Distillation yields 15.4 g, 88%, of 3-ethyl-1-pentanol, bp 169-170° (740 mm), n^{20}D 1.4296.

Notes

1. 3-Ethyl-2-pentene, bp 94-95°, n^{20}D 1.4148, was prepared by dehydration of 3-ethyl-3-pentanol with iodine. It may be dried as in **S9.11**.
2. 3-Ethyl-2-pentene undergoes hydroboration to the dialkylborane stage. However, after isomerization takes place to the terminal position, a third mole of olefin is taken up. Therefore, the reaction is carried out with a ratio of 3 olefin/BH_3, with the borane in modest excess.
3. Boron trifluoride diglymate (**S9.11**) was used to avoid the necessity of removing ethyl ether during the heating process.

[1] H. C. Brown and G. Zweifel, *J. Amer. Chem. Soc.*, 88, 1433 (1966).

6.2. 3-ETHYL-1-PENTENE

$$8\,(C_2H_5)_2C{=}CHCH_3 + 3\,NaBH_4 + 4\,BF_3 \xrightarrow[25^\circ]{DG} 4[(C_2H_5)_2CHCH(CH_3)]_2BH + 3\,NaBF_4$$

$$[(C_2H_5)_2CHCH(CH_3)]_2BH + (C_2H_5)_2CH{=}CHCH_3 \xrightarrow[160^\circ,\ 2\ hr]{DG} [(C_2H_5)_2CHCH_2CH_2]_3B$$

$$[(C_2H_5)_2CHCH_2CH_2]_3B + 3\,CH_3(CH_2)_7CH{=}CH_2 \xrightarrow[160^\circ]{DG} 3(C_2H_5)_2CHCH{=}CH_2 + [CH_3(CH_2)_9]_3B$$

Procedure by M. V. Bhatt[1]

Procedure

The procedure is carried out as described in **P6.1** up to and including the isomerization step. Then the apparatus is connected to an efficient fractionating column (**N1**) with a flow of nitrogen to protect the hydroboration mixture. To the reaction mixture is added 31.5 g [(0.225 mole), 50% excess] of 1-decene (**S9.11**), and the mixture is heated under reflux as the 3-ethyl-1-pentene distills. The reaction is essentially complete in 6 hours, but the yield can be modestly improved by continuing the displacement for an additional period. There is obtained 12.0 g (82% yield) of 3-ethyl-1-pentene. VPC analysis revealed the product to contain 2% of 3-ethyl-2-pentene.

Notes

1. In the original procedure a Todd microcolumn was used. However, any column should be satisfactory which is capable of separating quantitatively the displaced olefin from the displacing olefin (3-ethyl-1-pentene from 1-decene in the present case).

[1] H. C. Brown and M. V. Bhatt, *J. Amer. Chem. Soc.*, 88, 1440 (1966).

6.3. 2,2,4-TRIMETHYL-1,5-PENTANEDIOL

$$(CH_3)_3CCH_2C(CH_3)=CH_2 \xrightarrow{\text{(thexyl)}BH_2} (CH_3)_3CCH_2CH(CH_3)CH_2B(H)\text{-thexyl}$$

$$(CH_3)_3CCH_2CH(CH_3)CH_2B(H)\text{-thexyl} \xrightarrow{200^\circ} \text{cyclic borane} + H_2$$

H3C, CH2, CH3; H3C C CH2; C H; CH2; B

$$\text{cyclic borane} \xrightarrow[50^\circ]{NaOH,\ H_2O_2} H_2C(OH)-C(CH_3)_2-CH_2-CH(CH_3)-CH_2OH$$

Procedure by Kenneth J. Murray[1]

Procedure

Thexylborane (0.100 mole) is prepared in tetrahydrofuran **(P2.10)**. The nitrogen atmosphere is maintained until after the final oxidation. To the thexylborane at 0° is added 11.2 g (0.100 mole) of 2,4,4-trimethyl-1-pentene **(N1)** over 10 minutes. The mixture is stirred for 1 hour. Then the tetrahydrofuran is distilled off through a short Vigreux column **(N2)**, and the temperature taken up to 200°. The mixture is heated for 10 hours. A total of 2.5 ℓ (0.100 mole) of hydrogen is evolved **(N3)**. The reaction mixture is cooled to room temperature, 50 ml of tetrahydrofuran is added, and the product is treated with 33 ml of 3*M* sodium hydroxide, followed by the slow, dropwise addition of 33 ml of 30% hydrogen peroxide, maintaining the temperature at 50° **(P2.1:N6)**. The aqueous phase is saturated with potassium carbonate **(P2.4:N3)**, and the tetrahydrofuran layer is separated. The aqueous phase is further extracted with tetrahydrofuran (2 × 25 ml), and the combined tetrahydrofuran phase is dried over magnesium sulfate. VPC examination revealed a yield of 81% **(N4)**. Distillation yielded 10.3 g (71% yield) of 2,2,4-trimethyl-1,5-pentanediol, bp 132-133° (8.5 mm), n^{20}D 1.4565.

Notes

1. Dried over a small quantity of lithium aluminum hydride, n^{20}D 1.4093 **(S9.11)**.
2. It is convenient to have the Vigreux column attached prior to the synthesis of the thexylborane.
3. The hydrogen should be conducted away and safely vented.
4. The VPC analysis revealed the presence of 0.090 mole of 2,3-dimethyl-1-butanol. The boron atom migrates from the tertiary position of the thexyl unit to the primary position during the thermal treatment.

[1]H. C. Brown, K. J. Murray, H. Müller, and G. Zweifel, *J. Amer. Chem. Soc.*, 88, 1443 (1966).

6.4. (−)-*CIS*- AND (−)-*TRANS*-PINANE

12 (CH_2=) + 3 $NaBH_4$ + 4 BF_3 $\xrightarrow[25°]{DG}$ (CH_2−$)_3$B + 3 $NaBF_4$

(CH_2−$)_3$B $\xrightarrow[2.5\ hr]{125°}$ (CH_2−$)_3$B

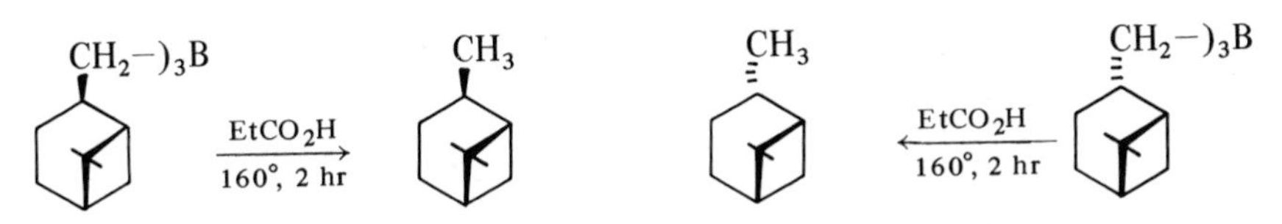

Procedure by George Zweifel[1]

Procedure

A 200-ml flask is assembled with a magnetic stirring bar, an inlet with a rubber serum cap, a condenser, and a mercury bubbler. The apparatus is dried and a nitrogen atmosphere maintained through the protonation stage (**S9.1**). With the aid of a hypodermic syringe, there is added 13.6 g (0.100 mole) of (–)-β-pinene (**N1**), 25 ml of diglyme, and 30 ml of a 1.00*M* solution of sodium borohydride in diglyme (20% excess). To the rapidly stirred solution, cooled with an ice-water bath, is added 11.0 ml of a 3.65*M* solution of boron trifluoride in diglyme (**P6.1:N3**). After 1 hour, 11 ml of propionic acid (~50% excess) is added, and the mixture is heated (**N2**) under reflux (~160°) for 2 hours. The reaction mixture is cooled and sufficient 3*M* sodium hydroxide added to ensure an excess. The diglyme phase is diluted with an equal volume of pentane, separated, and washed with equal volumes of ice water five times to remove the diglyme. Evaporation of the pentane yielded 12.4 g of *cis*-pinane. The product, purified by preparative gas chromatography, exhibits n^{20}D 1.4624, α^{20}D –19.3° (1 dm).

For the preparation of the *trans* isomer, the procedure is identical through the hydroboration stage. The reaction mixture is heated at 125° for 2.5 hours to isomerize the product. To the reaction mixture is added 11 ml of propionic acid (~50% excess), and the reaction mixture is heated under reflux (~160°) for 2 hours. The hydrocarbon product was isolated as described above for the *cis* isomer. There is obtained 13.0 g of crude product. VPC examination revealed 98% *trans*- and 2% *cis*-pinane. The product, purified by preparative gas chromatography, exhibits n^{20}D 1.4610, α^{20}D –14.5° (1 dm).

Notes

1. Commercial, n^{20}D 1.4794, $[\alpha]^{20}$D –21.1° (**S9.11**).
2. Sodium borohydride is soluble in diglyme at 25°, but precipitates as a diglymate at 0°. On the other hand, sodium fluoborate is soluble in diglyme at 0°, but separates as the temperature is raised.
3. Hydroboration-protonolysis of (+)-α-pinene,[1] $[\alpha]^{20}$D +47.6, yields (+)-*cis*-pinane, n^{20}D 1.4614, α^{20}D +21.5° (1 dm). Isomerization-protonolysis of the hydroboration product from (+)-α-pinene should yield (+)-*trans*-pinane.

[1] G. Zweifel and H. C. Brown, *J. Amer. Chem. Soc.*, **86**, 393 (1964).

6.5. *CIS*-STILBENE

$$C_6H_5C{\equiv}CC_6H_5 + Sia_2BH \xrightarrow[-10^\circ \text{ to } 0^\circ]{THF} (C_6H_5)(H)C{=}C(C_6H_5)(BSia_2)$$

$$(C_6H_5)(H)C{=}C(C_6H_5)(BSia_2) + HOAc \xrightarrow[25^\circ,\ 2\ hr]{THF} (C_6H_5)(H)C{=}C(C_6H_5)(H) + Sia_2BOAc$$

Procedure by George Zweifel[1]

Procedure

Disiamylborane (0.150 mole) is prepared in tetrahydrofuran (**P2.9**). The nitrogen atmosphere is maintained until following the protonolysis (**S9.1**). To the rapidly stirred solution cooled to −10° is added 26.7 g (0.150 mole) of diphenylacetylene (**N1, N2**). The solution is brought to room temperature and allowed to remain there for 2 hours to ensure completion of the hydroboration. Protonolysis is accomplished by adding 75 ml of anhydrous acetic acid and allowing the mixture to stir for 2 hours at room temperature (**N3**). The entire mixture is poured into ice water and taken up in ether. The disiamylborinic acid in the ether-tetrahydrofuran mixture is oxidized by the addition of 50 ml of 3*N* sodium hydroxide and 33 ml of 30% hydrogen peroxide (**N4**). After extraction and drying, distillation yields 18.6 g (69% yield) of *cis*-stilbene, bp 138-139° at 10 mm, n^{20}D 1.6212.

Notes

1. Commercial: mp 59-61°.
2. In order to minimize dihydroboration, it is usually preferable to add the hydroboration agent to the acetylene. In the present case, dihydroboration is sufficiently slow that the rapid addition of diphenylacetylene to the cooled solution of disiamylborane can be utilized for its convenience in avoiding the transfer of the disiamylborane solution.
3. In some cases, the protonolysis requires more drastic conditions. For example, a 95% yield of *cis*-propenylcyclohexane was achieved by protonolysis at 100°.[2]

$$\text{C}_6\text{H}_{11}\text{CH=C(CH}_3\text{)B(O}_2\text{C}_6\text{H}_4) \xrightarrow[2\ \text{hr}]{100°} \text{C}_6\text{H}_{11}\text{CH=CHCH}_3 \quad 95\%$$

Even more severe conditions are required for the intermediate in the Zweifel synthesis of *trans*-olefins[3] (**P8.18**).

$$\text{R}^1\text{C(B(OCH}_3\text{)}(\text{thexyl}))\text{=CHR}^2 \xrightarrow[\text{or EtCO}_2\text{H, 4 hr reflux}]{\text{HOAc, 24 hr reflux}} \text{R}^1\text{HC=CHR}^2$$

4. In many cases this oxidation step can be avoided. However, the volatilities of the boron by-product and *cis*-stilbene are so similar that difficulty is encountered in attempted separation by simple distillation.

[1] H. C. Brown and G. Zweifel, *J. Amer. Chem. Soc.*, **83**, 3834 (1961).
[2] H. C. Brown and S. K. Gupta, *J. Amer. Chem. Soc.*, **94**, 4370 (1972).
[3] E. Negishi, J. -J. Katz, and H. C. Brown, *Synthesis*, 555 (1972).

6.6. *CIS*-MYRTANYL IODIDE

$$3\ (\beta\text{-pinene}) + \text{H}_3\text{B:THF} \xrightarrow[25°]{\text{THF}} (\text{myrtanyl-CH}_2-)_3\text{B}$$

$$(\text{myrtanyl-CH}_2)_3\text{B} + \text{I}_2 + \text{NaOH(CH}_3\text{OH)} \xrightarrow[25°]{\text{THF}} 2\ \text{myrtanyl-CH}_2\text{I} + \text{myrtanyl-CH}_2\text{B(OH)}_2$$

Procedure by Michael W. Rathke and Milorad M. Rogić[1]

Procedure

A 500-ml flask equipped with a septum inlet, magnetic stirring bar, pressure-equalizing dropping funnel, and mercury bubbler is assembled, flushed with nitrogen, and maintained under a static nitrogen pressure in the usual manner (**S9.1**). The flask is charged with 75 ml of tetrahydrofuran and 20.4 g (23.4 ml, 0.100 mole) of (-)-β-pinene (**P6.4:N1**), and placed in a water bath at 25°, Conversion to the trialkylborane is achieved by the dropwise addition of 25.8 ml of a 2.00*M* solution of borane (0.052 mole) in tetrahydrofuran (**P2.3**) (**P2.4:N3**) to the stirred solution. After 1 hour at 25°, 1 ml of methanol is added to destroy the minor amounts of excess hydride. Then 28.0 g (0.110 mole) of iodine is added all at once, followed by the dropwise addition of 50 ml of a 3*M* solution of sodium hydroxide in methanol (0.150 mole) over 5 minutes. VPC analysis of the reaction mixture indicates the formation of 60% *cis*-myrtanyl iodide. The reaction mixture is poured into 150 ml of water containing 3 g of sodium thiosulfate to remove excess iodine, and the aqueous layer is extracted with two 100-ml portions of pentane. The combined pentane layers are dried and distilled. There is obtained 16.0 g (40% yield) of *cis*-myrtanyl iodide, bp 85° (0.8 mm), n^{20}D 1.5467 $[\alpha]^{26}$D − 44.6° (**N1, N2**).

Notes

1. Presumably, treatment of the thermally isomerized trimyrtanylborane (**P6.4**) would provide the isomeric *trans*-myrtanyl iodide.
2. In cases where disiamylborane is utilized as an intermediate it proved desirable to oxidize the disiamylboronic acid prior to isolation of the product (**P6.5**).

$CH{=}CH_2$ (cyclohexene) ⟶ $CH_2CH_2BSia_2$ (cyclohexene) ⟶ CH_2CH_2I (cyclohexene) + Sia_2BOH

[1] H. C. Brown, M. W. Rathke, and M. M. Rogić, *J. Amer. Chem. Soc.*, **90**, 5038 (1968).

6.7. METHYL 11-BROMOUNDECANOATE

$$3\ CH_3O_2C(CH_2)_8CH{=}CH_2 + H_3B{:}THF \xrightarrow[0^\circ]{THF} [CH_3O_2C(CH_2)_{10}]_3B$$

$$CH_3O_2C(CH_2)_{10}]_3B + 3\,Br_2 + 4\,NaOCH_3 \xrightarrow[0^\circ]{THF}$$

$$3\,CH_3O_2C(CH_2)_{10}Br + 3\,NaBr + NaB(OCH_3)_4$$

Procedure by Clinton F. Lane[1]

Procedure

A dry 500-ml flask equipped with a magnetic stirring bar, septum inlet, thermometer, pressure-equalizing dropping funnel, and mercury bubbler is flushed with nitrogen and then maintained under a static nitrogen pressure (**S9.1**). The flask is charged with 75 ml of dry tetrahydrofuran (**S9.11**) and 29.7 g (33.5 ml, 0.150 mole) of dry methyl 10-undecenoate (**N1**) and then cooled to ~0° with an ice-water bath. Hydroboration is achieved by the dropwise addition of 25.2 ml of 2.00*M* borane in tetrahydrofuran (**P2.3**) (**P2.4:N3**) (**N2**). The solution is stirred for 0.5 hour at 0° and 0.5 hour at 20°. Then 1 ml of methanol is added to destroy traces of residual hydride. The solution is cooled to below 0° with an ice-salt bath, and 10.0 ml (0.20 mole) of bromine is added at such a rate that the temperature of the reaction mixture does not rise above 0°. Then 60 ml of a 4.16*M* solution of sodium methoxide (0.250 mole) in methanol (**N3**) is added dropwise over a period of 45 minutes. The temperature is maintained below 5° during the addition of the base. The reaction flask is then placed in a water bath at 20-25°, and the reaction mixture is treated with 50 ml of pentane, 20 ml of water, and 20 ml of saturated aqueous potassium carbonate. The organic layer is separated from the aqueous layer, and the aqueous layer is extracted with pentane (3 × 50 ml). The pentane extracts are combined with the organic layer, washed twice with water (50 ml), and once with 50 ml of saturated aqueous sodium chloride. After drying the solution over anhydrous potassium carbonate, the pentane is removed on a rotary evaporator under reduced pressure, providing 41.2 g, 93% of a colorless oil (**N4**). Distillation gives 35.4 g (85% yield) of methyl 11-bromoundecanoate, bp 126-128° (0.65 mm), n^{20}D 1.4638 (**N5, N6**).

Notes

1. Commercial: bp 123-124° (12.5 mm), n^{20}D 1.4392 (**S9.11**).
2. Excess borane should be minimized in order to avoid appreciable concurrent reduction of the ester group.
3. The sodium methoxide solution should be prepared by slowly adding methanol to sodium metal. (Commercial sodium methoxide was less satisfactory.)
4. VPC examination revealed the presence of approximately 1% of another peak, assumed to be the 10-bromo isomer. Since the hydroboration product

must contain approximately 6% of the secondary isomer, the reaction must utilize the primary groups selectively.

5. Commercial BH_3:THF provided a lower yield and a less pure product.
6. An alternative procedure was also developed involving the simultaneous addition at 25° of the stoichiometric amount of bromine and 10% excess sodium methoxide. This procedure proved somewhat more advantageous for secondary alkyl derivatives.[1]

[1] H. C. Brown and C. F. Lane, *J. Amer. Chem. Soc.*, **92**, 6660 (1970).

6.8. *TRANS*-1-OCTENYL IODIDE

$$n\text{-}C_6H_{13}C{\equiv}CH + \text{catecholborane (C}_6H_4O_2BH) \xrightarrow[70^\circ]{\text{neat}} trans\text{-}n\text{-}C_6H_{13}CH{=}CH\text{-}B(O_2C_6H_4)$$

$$trans\text{-}n\text{-}C_6H_{13}CH{=}CH\text{-}B(O_2C_6H_4) + 2\,H_2O \xrightarrow[25^\circ]{H_2O} trans\text{-}n\text{-}C_6H_{13}CH{=}CH\text{-}B(OH)_2 + C_6H_4(OH)_2$$

$$trans\text{-}n\text{-}C_6H_{13}CH{=}CH\text{-}B(OH)_2 + NaOH + I_2 \xrightarrow[0^\circ]{EE} trans\text{-}n\text{-}C_6H_{13}CH{=}CHI + NaI + NaB(OH)_4$$

Procedure by Tsutomu Hamaoka and Nair Ravindran[1]

Procedure

The synthesis of 0.100 mole of 2-(*trans*-1-octenyl)-1,3,2-benzodioxaborole from 1-octyne (**S9.11**) and catecholborane follows precisely the procedure of **P4.11**. After the 2-hour reaction period at 70°, the flask and its contents are cooled to 25°, and 100 ml of water is added. The mixture is stirred for 2 hours at

25° to effect hydrolysis of the ester. The mixture is cooled to 0°, and the white solid, *trans*-1-octenylboronic acid, is collected by filtration (**N1**). It is washed several times on the filter with ice-cold water to remove the last traces of catechol. The yield of the boronic acid is 90%.

The *trans*-1-octenylboronic acid prepared above is dissolved in 100 ml of ether in a 1-ℓ flask and cooled to 0° in an ice bath. To the stirred solution is added 100 ml of 3*M* sodium hydroxide followed by 30.5 g (0.120 mole) of iodine in 300 ml of ether. The stirring is continued for 0.5 hour following completion of the addition of the iodine. A few drops of a saturated solution of sodium thiosulfate in water is then added to destroy residual iodine, as indicated by the disappearance of the color. The ether solution is separated from the aqueous phase and dried over anhydrous magnesium sulfate. VPC analysis indicates a yield of 100% based on boronic acid (90% based on 1-octyne). The ether is removed with a water aspirator, and the product is distilled: 16.9 g (71% yield based on 1-octyne), bp 58° (0.2 mm). The isomeric purity is greater than 99% *trans* (**N2, N3**).

Notes

1. The trans-1-octenylboronic acid is stable to air, and its subsequent conversion to iodide can be carried out open to the atmosphere.
2. Other terminal acetylenes, such as cyclohexylethyne and phenylethyne, can be converted to the *trans*-iodides similarly. However, the reaction apparently takes another course with internal acetylenes.[1]
3. 1-Octyn-3-ol, protected with the tetrahydropyranyl group, has been hydroborated with disiamylborane and oxidized to the boronic acid with trimethylamine oxide. This boronic acid was also converted into the corresponding *trans* iodide by treatment with base and a large excess of iodine.[2]

[1] H. C. Brown, T. Hamaoka, and N. Ravindran, *J. Amer. Chem. Soc.*, **95**, 5786 (1973).
[2] A. F. Kluge, K. G. Untch, and J. H. Fried, *J. Amer. Chem. Soc.*, **94**, 7827 (1972).

6.9. *CIS*-1-OCTENYL BROMIDE

$n\text{-}C_6H_{13}C\equiv CH$ + (catecholborane, $C_6H_4O_2BH$) $\xrightarrow[70^\circ]{\text{neat}}$ $n\text{-}C_6H_{13}(H)C{=}C(H)B(O_2C_6H_4)$

$$n\text{-}C_6H_{13}CH{=}CH\text{-}B(O_2C_6H_4) + 2\,Br_2 \xrightarrow[0^\circ]{CH_2Cl_2} n\text{-}C_6H_{13}CHBr\text{-}CHBr\text{-}B(O_2C_6H_3Br) + HBr$$

$$n\text{-}C_6H_{13}CHBr\text{-}CHBr\text{-}B(O_2C_6H_3Br) + 2\,NaOCH_3 \xrightarrow[0^\circ]{CH_3OH} n\text{-}C_6H_{13}CH{=}CHBr + [BrC_6H_3O_2B(OCH_3)_2]^- Na^+ + NaBr$$

Procedure by Tsutomu Hamaoka and Nair Ravindran[1]

Procedure

The 2-(*trans*-1-octenyl)-1,3,2-benzodioxaborole is synthesized from 0.100 mole each of 1-octyne and catecholborane (**P6.8, P4.11**) at 70°. The flask is cooled to room temperature, and 125 ml of methylene chloride is added (under nitrogen). The flask is immersed in an ice bath and the temperature reduced to approximately 0° (**N1**). Then 32.0 g (0.200 mole) of bromine is added with stirring, maintaining the temperature below 5°. After 1 hour, 200 ml of 2*M* sodium methoxide (**P6.7:N3**) in methanol is added and the mixture stirred at 0°. The reaction mixture is now brought to room temperature, 100 ml of water added, and the organic phase is separated (**N2**). The aqueous phase is extracted twice with methylene chloride (2 X 25 ml), and the combined organic phase is dried over anhydrous magnesium sulfate. Distillation yields 15.6 g (82% yield based on 1-octyne, 91% on the intermediate) of *cis*-1-octenyl bromide, bp 90-91° (35 mm), n^{20}D 1.4619. VPC examination establishes the isomeric purity to be 99% *cis*.

Notes

1. The particular temperature required to achieve stereospecificity of approximately 99% varies with the structure of the acetylene. Thus the derivative from phenylacetylene requires a temperature of −40° for the bromine addition.[1]

2. The separation and subsequent operations can be carried out open to the atmosphere.
3. The reaction is also applicable to internal acetylenes.[1]

$$C_2H_5C{\equiv}CC_2H_5 \longrightarrow (C_2H_5)(H)C{=}C(Br)(C_2H_5)$$

$$(CH_3)_3CC{\equiv}CCH_3 \longrightarrow ((CH_3)_3C)(H)C{=}C(Br)(CH_3)$$

[1] H. C. Brown, T. Hamaoka, and N. Ravindran, *J. Amer. Chem. Soc.*, **95**, 6456 (1973).

6.10. THEXYLCYCLOHEXYLCYCLOPENTYLCARBINOL

$$\text{Thexyl(cyclopentyl)(cyclohexyl)C–B}\langle O\text{–}CH_2\text{–}CH_2\text{–}O\rangle + H_2O_2 + NaOH \xrightarrow[50^\circ]{\text{THF-EtOH}} \text{Thexyl(cyclopentyl)(cyclohexyl)C–OH}$$

$$+ NaB(OH)_4 + (CH_2OH)_2$$

Procedure by Jean-Jacques Katz and Bruce A. Carlson[1]

Procedure

Treatment of thexylcyclohexylcyclopentylborane **(P4.18)** with α,α-dichloromethyl methyl ether and lithium triethylcarboxide yields an ester of thexylcyclohexylcyclopentylcarbinyl boronic acid, $RR'R''CB(OCH_3)Cl$ **(P8.9)**. The oxidation of this compound and related derivatives requires more forcing conditions than those required for the usual oxidation of organoboranes **(P2.1, P2.4) (N1)**. The above boronic acid ester (0.100 mole in 100 ml of tetrahydrofuran) is converted to the ethylene glycol derivative **(N2)** by adding 12.4 g (0.200 mole) of ethylene glycol and removing all the volatile materials by simple distillation. To the residue is added 50 ml of 95% ethanol and 25 ml of tetrahydrofuran, followed by 24 g (0.60 mole) of solid sodium hydroxide. When most of the sodium hydroxide has dissolved, 50 ml of 30% hydrogen peroxide is *cautiously* added dropwise to the solution over a period of 2 hours **(N3, N4)** maintaining the temperature below 50°. The reaction mixture is then maintained at 55-60° for an additional 2 hours. It is brought to 25°; the aqueous phase is saturated with sodium chloride, and the organic phase separated. The aqueous phase and salts are extracted with ether (3 × 50 ml), and the combined organic phase is dried over anhydrous magnesium sulfate. Following removal of the solvents, the triethylcarbinol is distilled at 50° (0.2 mm) and the product obtained as a very thick oil. In order to minimize dehydration, it was rapidly distilled in two portions in a Kügelrohr over to yield a total of 17.4 g (66% yield) of thexylcyclohexylcyclopentylcarbinol, bp 115-120° (0.5 mm), n^{20}D 1.5060.

Notes

1. In the usual oxidation of organoboranes, R_3B, R_2BOH, and $RB(OH)_2$, there is used approximately 1 mole of sodium hydroxide (as a 3*M* aqueous solution) per gram-atom of boron, and 1 mole of hydrogen peroxide (as the 30% aqueous solution, ~10*M*) per equivalent of carbon-boron bond. Thus the oxidation of 100 mmoles of R_3B requires 33.3 ml of 3*M* sodium hydroxide and 33.3 ml of 30% hydrogen peroxide **(P2.5)**. For the oxidation of 100 mmoles of R_2BOH, 33.3 ml of 3*M* sodium hydroxide and 22.2 ml of 30% hydrogen peroxide would be used **(P2.4)**. For the oxidation of 100 mmoles of $RB(OH)_2$, 33.3 ml of 3*M* sodium hydroxide and 11.1 ml of 30% hydrogen peroxide would be used.

   Most organoboranes oxidize quite vigorously, so the hydrogen peroxide must be added cautiously, dropwise, and the temperature maintained below 40-50° with external cooling. If the hydroboration medium is not miscible with water, such as ethyl ether **(P4.14:N1)** or hexane **(P2.6)**, it is desirable to add ethanol as a cosolvent.
2. Conversion to the ethylene glycol ester facilitates the oxidation. Evidently, formation of the cyclic ester makes the boron atom more available sterically to attack by the HO_2^- anion.
3. The present procedure utilizes roughly an equal volume of ethanol as a cosolvent, the use of excess sodium hydroxide in solid form (to minimize the amount of water), and the use of excess hydrogen peroxide. This procedure has handled every difficult case encountered but one.[2] CAUTION: Some molecular oxygen is generated when excess H_2O_2 is used. When this evolves from the reaction mixture, it often carries some solvent vapors with it. Consequently, this dangerous mixture of vapors should be vented in a fume hood, and all sources of open flame or sparks kept well away.
4. It is important that the solution be homogeneous or as nearly homogeneous as possible. The precipitation of the boronic ester or acid during the addition of the 30% hydrogen peroxide may result in incomplete oxidation. It is

therefore recommended that additional ethanol be added to dissolve any solid product which separates at 40-50°.

[1] H. C. Brown, J.-J. Katz, and B. A. Carlson, *J. Org. Chem.*, **38**, 3968 (1973).
[2] H. C. Brown and B. A. Carlson, *J. Organomet. Chem.*, **54**, 61 (1973).

6.11. 2-METHYLCYCLOHEXANONE

$$8\ \text{(1-methylcyclohexene)} + 3\,LiBH_4 + BF_3{:}OEt_2 \xrightarrow[25°]{EE} 4\ \text{(2-methylcyclohexyl)}_2BH + 3\,LiF$$

$$3\ \text{(2-methylcyclohexyl)}_2BH + H_2O \xrightarrow[25°]{EE} 6\ \text{(2-methylcyclohexyl)}_2BOH + H_2$$

$$3\ \text{(2-methylcyclohexyl)}_2BOH + 4\,Na_2Cr_2O_7 + 16\,H_2SO_4 \xrightarrow[25°]{EE} 6\ \text{(2-methylcyclohexanone)}$$

$$+ 4\,Cr_2(SO_4)_3 + 4\,Na_2SO_4 + 3(HO)_3B + 16\,H_2O$$

Procedure by Chandra P. Garg[1]

Procedure

The usual hydroboration apparatus is assembled using a 200-ml flask, magnetic stirrer, condenser, pressure-equalizing addition funnel, thermometer, and bubbler (**S9.1**). The system is flushed with nitrogen and maintained under a static pressure of nitrogen through the oxidation stage. In the flask are placed 0.5 g (22.5 mmole, 20% excess) of lithium borohydride (**P4.12:N2**) in 30 ml of ethyl ether (**N1, N2**) and 4.8 g (0.050 mole) of 1-methylcyclohexene (**S9.11**). The flask is immersed in a water bath at 25°. To the flask is added 0.95 ml (0.0075 mole, 20% excess) of boron trifluoride etherate over a period of 15 minutes at 25-30°. After 2 hours, 5 ml of water is added to destroy excess residual hydride. The chromic acid solution (10% excess), prepared from 11.0 g (39.6 mmole) of sodium dichromate dihydrate and 8.25 ml (148.3 mmole) of 96% sulfuric acid and diluted with water to 45 ml, is added to the stirred solution over a period of 15 minutes, maintaining the temperature at 25-30°. The mixture is then heated under reflux for 2 hours. The upper layer is separated, and the aqueous phase is extracted with two 10-ml portions of ether. VPC examination indicates an 87% yield of 2-methylcyclohexanone. Distillation gives 4.36 g (78% yield) of 2-methylcyclohexanone, bp 63-64° (24 mm), n^{20}D 1.4487 (**N3**).

Notes

1. This small scale reaction utilizes ethyl ether because this solvent greatly facilitates separation of the product from the inorganic salts and permits ready recovery of the product.
2. Attention is called to the special convenience ethyl ether provides for a two-phase oxidation of secondary alcohols to ketones with aqueous chromic acid.[2]
3. On a much larger scale organoboranes, synthesized in diglyme or in tetrahydrofuran, have been successfully oxidized to ketones by aqueous chromic acid[3] (**S5.8**).

[1] H. C. Brown and C. P. Garg, *J. Amer. Chem. Soc.*, **83**, 2951 (1961).
[2] H. C. Brown, C. P. Garg, and K.-T. Liu, *J. Org. Chem.*, **36**, 387 (1971).
[3] H. C. Brown, I. Rothberg, and D. L. Vander Jagt, *J. Org. Chem.*, **37**, 4098 (1972).

6.12. CYCLOHEXYL HYDROPEROXIDE

$$3\ \text{cyclohexene} + H_3B{:}THF \xrightarrow[50^\circ]{THF} (\text{cyclohexyl})_3B$$

$$(\text{cyclohexyl})_3B + 2\,O_2 \xrightarrow[-78^\circ]{THF} (\text{cyclohexyl}O_2)_2B\text{-cyclohexyl}$$

$$(\text{cyclohexyl}O_2)_2B\text{-cyclohexyl} + H_2O_2 \xrightarrow[0^\circ]{THF} 2\ \text{cyclohexyl}O_2H + \text{cyclohexyl}OH + (HO)_3B$$

Procedure by M. Mark Midland[1]

Procedure

A dry 200-ml flask equipped with a septum inlet and a magnetic stirrer (fitted with a Teflon collar, **N1**) is flushed with nitrogen and maintained under a static nitrogen pressure (**S9.1**). The flask is charged with 75 ml of dry tetrahydrofuran

(S9.11) and 12.3 g (0.150 mole) of cyclohexene **(S9.11)**, and the flask is immersed in an ice-water bath. Hydroboration is achieved by the dropwise addition of 25.0 ml of 2.00*M* borane in tetrahydrofuran **(P2.3) (N2)**, followed by heating at 50° for 3 hours to ensure completion of this relatively sluggish hydroboration **(N3)**.

The automatic gasimeter **(S9.4)** is assembled for oxygen generation with standardized 15% hydrogen peroxide in the buret and basic manganese dioxide in the generator flask to convert the hydrogen peroxide to oxygen. An empty 100-ml flask is temporarily attached, and the automatic gasimeter is thoroughly flushed with oxygen by slowly injecting 15 ml of 30% hydrogen peroxide into the generator.

The flask containing the tricyclohexylborane is cooled (under the nitrogen atmosphere) to −78°. The temporary 100-ml flask is removed from the automatic gasimeter and quickly replaced by the −78° flask containing the tricyclohexylborane (the nitrogen blanket and lack of stirring inhibit the absorption of oxygen from the generator). The nitrogen blanket is removed by the controlled injection of 5 ml of 30% of hydrogen peroxide, and vigorous stirring is initiated. Oxygen absorption is followed quantitatively from the amount of standard hydrogen peroxide in the buret that is utilized by the system. In this case the absorption of oxygen per mole of tricyclohexylborane approaches 2 moles (approximately 40 ml of 15% hydrogen peroxide). The reaction mixture is brought to 0° to complete the absorption **(N4)**. To the solution at 0° is added dropwise 16.5 ml of 30% aqueous hydrogen peroxide. The solution is stirred at 0°. for 0.5 hour. Hexane, 50 ml, is added and the solution washed with 2 × 25 ml of water. Iodometric titration for hydroperoxide[3] inidicated a yield of 95%.

The tetrahydrofuran-hexane solution is extracted four times with 25-ml portions of 40% aqueous potassium hydroxide. The combined aqueous extracts, containing the hydroperoxide as the salt, is washed with 50 ml of hexane. The aqueous extract is then carefully neutralized at 0° with concentrated hydrochloric acid to liberate the hydroperoxide, which is taken up in 2 × 50 ml of hexane. The hexane solution is dried over anhydrous magnesium sulfate and distilled. There is obtained 9.5 g (82% yield) of cyclohexyl hydroperoxide, bp 39-40° (0.08 mm), n^{20}D 1.4645 (lit.[4] bp 42-43° (0.1 mm), n^{20}D 1.4645).

Notes

1. The Teflon collar permits a more vigorous stirring.[2] It is important to keep the solution saturated with oxygen to achieve the desired rapid conversion to the peroxide.
2. CAUTION! The oxidized solution from commercial BH_3:THF, containing $NaBH_4$ stabilizer, gave an exothermic reaction on warming from −78° with lower yields (~40%).
3. Alternatively, the reaction can be completed in less than 1 hour at reflux. In the case of most olefins, hydroboration proceeds to completion in a few

minutes at 25° or even 0°. However, dicyclohexylborane is of low solubility **(P2.8)**, and cyclohexene is an olefin of low reactivity.

4. This procedure appears to be general for organoboranes from internal olefins (i.e., R_3B with secondary alkyl groups). In the case of organoboranes from terminal olefins, the absorption of only 1 mole of oxygen proceeds rapidly at -78°. The absorption of the second mole then takes place at 0°. Otherwise, the procedures are identical. The reaction mixtures should not be permitted to warm up until the indicated amount of oxygen has been absorbed.

[1] H. C. Brown and M. M. Midland, *J. Amer. Chem. Soc.*, 93, 4078 (1971).
[2] C. A. Brown and H. C. Brown, *J. Amer. Chem. Soc.*, 84, 2829 (1962).
[3] S. Siggia, *Quantitative Analysis via Functional Groups,* John Wiley and Sons, New York, 1949.
[4] C. Walling and S. A. Buckler, *J. Amer. Chem. Soc.*, 77, 6032 (1955).

6.13. METHYL *EXO*-NORBORNYL SULFIDE

3 (norbornene) + $H_3B{:}THF$ $\xrightarrow[0°]{THF}$ (norbornyl$)_3B$

(norbornyl$)_3B$ + 2 CH_3SSCH_3 $\xrightarrow[THF,\ 25°]{cat.\ O_2}$ 2 (norbornyl)$-SCH_3$ + (norbornyl)$-B(SCH_3)_2$

Procedure by M. Mark Midland[1]

Procedure

A dry 200-ml flask equipped with a magnetic stirring bar, a septum inlet, and a connector to a mercury bubbler is flushed with nitrogen and maintained under a static pressure of nitrogen **(S9.1)**. The flask is charged with 80 ml of dry tetrahydrofuran **(P9.11)** and 14.4 g (0.150 mole) of norbornene and cooled to 0°. Hydroboration is achieved by the dropwise addition of 25.0 ml of a 2.00*M* solution of borane (0.050 mole) in tetrahydrofuran **(P2.3) (P2.4:N3)** at 0°, followed by stirring at room temperature for 1 hour. Methyl disulfide, 9.4 g (0.100 mole), is added to the stirred solution and the reaction initiated by introducing air **(N1)** at a rate of 1 ml/minute through a syringe needle penetrating the spectum inlet to a point just above the tetrahydrofuran solution **(N2)**. The reaction is allowed to proceed for 10 hours and then distilled using an efficient (Widmer) column **(S9.7)**. There is obtained methyl *exo*-norbornyl sulfide **(N3)**, 10.6 g (74% yield), bp 90-92° (23 mm), n^{20}D 1.5115, sulfone mp 74.5-75° [lit.[2] bp 82° (18 mm), n^{20}D 1.5118, sulfone mp 75°] **(N4)**.

Notes

1. The reaction can also be initiated photochemically with higher yields. In this procedure, the apparatus includes a reflux condenser. Following hydroboration of the olefin, the tetrahydrofuran is removed with a water aspirator and replaced with an equal volume of hexane. The methyl disulfide is then added, and the solution is illuminated with an external ultraviolet sunlamp (the commercially available Sears 275-W sunlamp was used). The hexane is allowed to reflux. The reaction is complete in 2 hours. The VPC yield of methyl *exo*-norbornyl sulfide is 95%.
2. Some free radical attack on the tetrahydrofuran occurs, with the formation of methyl 2-tetrahydrofuryl sulfide. This can be avoided and the yields improved, as in the photochemical reaction, by replacing the tetrahydrofuran with a less reactive solvent, such as hexane **(N1)**.
3. The product appeared to be homogeneous by VPC and NMR examination. The *exo* assignment rests on the strong preference for *exo* attack in the norbornyl system and the melting point of the sulfone.
4. The reaction uses only two of the three norbornyl groups on boron. More efficient use of the alkyl group can be achieved by use of 3,5-dimethylborinane **(P2.12, P8.24)**.

B H → B; $h\nu$, CH_3SSCH_3 → SCH_3 94%

[1]H. C. Brown and M. M. Midland, *J. Amer. Chem. Soc.*, **93**, 3291 (1971).
[2]D. I. Davies, L. T. Parfitt, C. K. Alden, and J. A. Claisse, *J. Chem. Soc. C*, 1585 (1969).

6.14. (-)-*CIS*-MYRTANYLAMINE

3 + $H_3B{:}THF$ $\xrightarrow[0°]{THF}$ $(CH_2)_3B$

$(CH_2)_3B$ + 2 H_2NOSO_3H $\xrightarrow[\text{reflux}]{\text{THF}}$ CH_2-NH_2

Procedure by Clinton F. Lane[1]

Procedure

A 500-ml three-neck flask fitted with a septum inlet, reflux condenser, pressure equalized addition funnel, a magnetic stirring bar, and a mercury bubbler is oven dried, assembled hot, and flushed with nitrogen, then maintained under a static pressure of the gas **(S9.1)**. Dry (-)-β-pinene (0.18 mole, 24.5 g, 28.5 ml) **(N1)** is added through the septum inlet followed by 200 ml of dry THF **(S9.11)**. The flask is cooled in an ice-water bath, and hydroboration is achieved by the dropwise addition of 0.063 mole BH_3:THF (35.2 ml of 1.79 *M*, 5% excess) from the addition funnel to the stirred solution. After the addition is complete, the mixture is stirred for 1 hour at room temperature. In a dry, nitrogen flushed 100-ml flask fitted with septum inlet, 15 g of hydroxylamine-0-sulfonic acid **(N2)** is slurried with 50 ml of dry THF. The slurry is transferred to the main reaction flask through the septum inlet using a large-bore double-ended needle. The reaction mixture soon begins to boil. The reflux is maintained by external heating. A 50-ml second portion of THF is used to slurry any remaining hydroxylamine-0-sulfonic acid in the 100-ml flask, and this is transferred as before into the main flask. The reaction mixture is heated at reflux for 3 hours and then allowed to stand overnight. The reaction mixture is poured into a mixture of 80 ml of concentrated hydrochloric acid and 500 g of ice. The water layer is extracted with ether (3 X 200 ml) **(N3)**. The extracts are discarded, and the aqueous layer is cooled in an ice-water bath. The acidic solution is treated with solid sodium hydroxide pellets until a pH of 10 is obtained, then it is extracted with ether (3 X 100 ml). The combined extracts are dried over potassium carbonate, filtered, concentrated on a rotary evaporator, and distilled to give 13.4 g (48%) of (-)-*cis*-myrtanyl amine, bp 94-99° (27 mm), VPC 99% pure, $n^{20}D$ 1.4877 $[a]^{22}D$ −30.5° (neat) **(N4)**.

Notes

1. Commercial, bp 165-167°, $n^{20}D$ 1.4782, $[a]^{20}D$ −21° (neat)**(S9.11)**.
2. Commercial hydroxylamine-0-sulfonic acid exhibited purities of 70-80%.[3] High purity material is available from Alfa Chemical Company. However, it is best prepared fresh by the procedure in *Inorganic Synthesis.*[3]
3. The reaction appears capable of utilizing two of the three groups in R_3B and one of the two alkyl groups in R_2BH. Consequently, the reaction product

contains an alkylboronic acid derivative which is removed by the extraction.

4. Even though the yields of amines are only in the range of 40 to 50%, this procedure provides a simple, one-stage synthesis of single epimeric amines for many structures.[2]

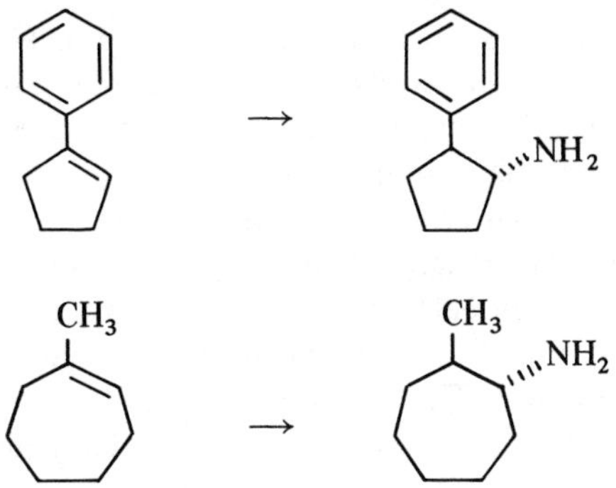

$\longrightarrow$ NH_2

CH_3 $\longrightarrow$ CH_3 NH_2

[1]Adapted from a procedure by H. C. Brown, W. R. Heydkamp, E. Breuer, and W. S. Murphy, *J. Amer. Chem. Soc.*, **86**, 3565 (1964).
[2]M. W. Rathke, N. Inoue, K. R. Varma, and H. C. Brown, *J. Amer. Chem. Soc.*, **88**, 2870 (1966).
[3]H. J. Matsuguma and L. F. Audrieth, *Inorg. Syn.*, **5**, 122 (1957).

6.15. *N-(TRANS*-2-METHYLCYCLOHEXYL)CYCLOHEXYLAMINE

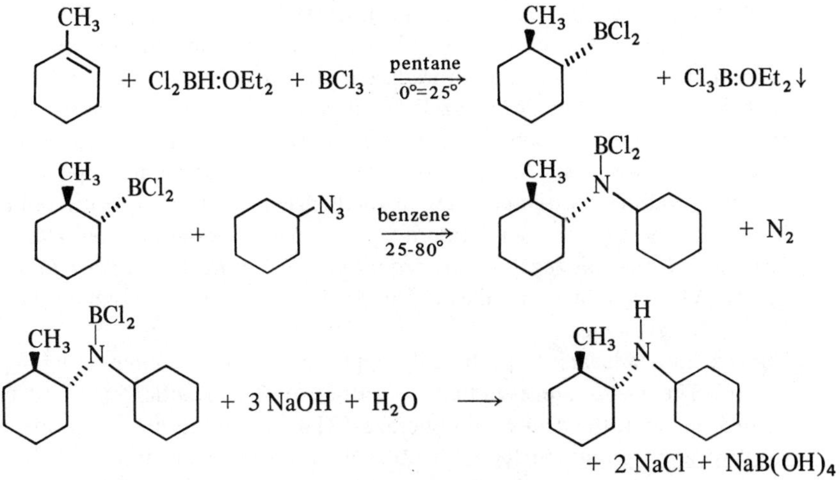

CH_3 + $Cl_2BH{:}OEt_2$ + BCl_3 $\xrightarrow[0°-25°]{\text{pentane}}$ CH_3 BCl_2 + $Cl_3B{:}OEt_2\downarrow$

CH_3 BCl_2 + N_3 $\xrightarrow[25\text{-}80°]{\text{benzene}}$ CH_3 BCl_2 N + N_2

BCl_2 CH_3 N + 3 NaOH + H_2O $\longrightarrow$ H CH_3 N + 2 NaCl + $NaB(OH)_4$

Procedure by M. Mark Midland and Alan B. Levy[1]

Procedure

A dry 500-ml flask equipped with a septum inlet, reflux condenser, magnetic stirring bar, and mercury bubbler is flushed with nitrogen and maintained under a static pressure of nitrogen until after the reaction with the azide (**S9.1**). The flask is charged with 100 ml of benzene and 17.8 g (0.100 mole) of *trans*-2-methylcyclohexyldichloroborane (**N1**). Then 12.5 g (0.100 mole) of cyclohexyl azide (**N2**) is added dropwise over 30 minutes at 20°. The solution is stirred an additional 15 minutes, and the temperature then brought to 80° (gentle refluxing of the solvent) over 45 minutes (**N3**). A gentle reflux is maintained until nitrogen evolution is complete. The solution is cooled to 0-5°, and the reaction product is carefully hydrolyzed by slowly adding 10 ml of water (exothermic!). An additional 90 ml of water is added and the precipitate removed by filtration. The organic phase is separated and washed with 50 ml of 2*M* hydrochloric acid. The combined aqueous layer and precipitate is made strongly alkaline with 40% potassium hydroxide. The amine is taken up in ether (4 × 25 ml), the ether extract dried over potassium carbonate, and the ether removed under vacuum. Distillation gives 16.8 g (86% yield) of *N*-(*trans*-2-methylcyclohexyl)cyclohexylamine, bp 102-104° (1.5 mm), n^{20}D 1.4790 (**N4, N5**).

Notes

1. *Trans*-2-methylcyclohexyldichloroborane, bp 120-122° (100 mm), is prepared in 90% yield by the reaction of dichloroborane with 1-methylcyclohexene following the procedure of **P4.13** except that it requires stirring for one hour at room temperature.
2. Cyclohexyl azide, bp 68-69° (21 mm), n^{20}D 1.4665, is prepared in 74% yield by the reaction of cyclohexyl bromide and sodium azide in dimethylformamide following a related procedure in the literature.[2]
3. With less hindered boranes or azides, such as *n*-butyl, the reaction is rapid, evolving nitrogen at room temperature. In the more hindered cases, the reaction is slower, requiring 3-4 hours for completion at room temperature. Gradually raising the temperature of the benzene solution to gentle reflux brings the reaction to completion in approximately 30-40 minutes. (*Rapid heating of large-scale reactions may result in a violent exothermic reaction!*).
4. Careful VPC examination indicates the product to be essentially pure *trans*, free of the *cis* isomer.
5. The reaction appears to be broadly applicable, even accommodating aryl groups. The yields are essentially quantitative. The reaction can also be carried out with the dialkylchloroboranes[2] (**P4.14**). In this case, only one of the two alkyl groups in R_2BCl is utilized. However, the synthesis of these derivatives is simpler, so the choice must be made on the basis of the relative

advantage of the procedure for the synthesis of the intermediate and the value of the group being introduced.

[1] H. C. Brown, M. M. Midland, and A. B. Levy, *J. Amer. Chem. Soc.*, **95**, 2394 (1973).
[2] A. J. Parker, *J. Chem. Soc.*, 1328 (1961).
[3] H. C. Brown, M. M. Midland, and A. B. Levy, *J. Amer. Chem. Soc.*, **94**, 2114 (1972).

6.16. *N*-CYCLOHEXYL-7-AZABICYCLO[4.1.0]HEPTANE

BCl_2 + I, N_3 $\xrightarrow[25\text{-}80^\circ, -N_2]{\text{benzene}}$ I, N, BCl_2

+ 3 KOH + H_2O ⟶ + 2 KCl + $KB(OH)_4$

+ KOH $\xrightarrow[80^\circ]{\text{benzene}}$ + KI + H_2O

Procedure by Alan B. Levy[1]

Procedure

A dry 50-ml flask equipped with a septum inlet, reflux condenser, magnetic stirring bar, and mercury bubbler is flushed with nitrogen, and a static nitrogen atmosphere is maintained until after the reaction with the azide is complete **(S9.1)**. The flask is charged with 10 ml of benzene and 1.65 g (10 mmoles) of cyclohexyldichloroborane **(N1)**. Then 2.50 g (10 mmoles) of *trans*-1-azido-2-iodocyclohexane **(N2)** is added dropwise. After the addition has been completed, the reaction mixture is brought to 80° over a period of 1 hour. The evolution of nitrogen is complete at this point **(P6.15:N3)**. (The nitrogen atmosphere is no longer required.) The solution is cooled to 0° and carefully hydrolyzed by the addition of 10 ml of 10% hydrochloric acid (exothermic reaction!). To ensure complete precipitation of the salt, 15 ml of hexane is added. The precipitate is removed by filtration. The organic phase is separated and washed with 30 ml of 10% hydrochloric acid. The combined aqueous layers and precipitate is made strongly basic with 40% potassium hydroxide. The amine is extracted with 30 ml of benzene and the resulting solution heated under reflux with 30 ml of 40% potassium hydroxide for 1 hour (**N3**). The organic phase is separated and dried over Drierite. Benzene is removed under vacuum. Distillation in a Kugelrohr oven gives 1.46 g (82% yield) of *N*-cyclohexyl-7-azabicyclo[4.1.0]heptane, bp 80-82° (41 mm), n^{20}D 1.4835 (**N4, N5**).

Notes

1. Cyclohexyldichloroborane, bp 166-168°, is prepared in 80% yield by the reaction of dichloroborane with cyclohexene following the procedure of **P4.13**.
2. *Trans*-1-azido-2-iodocyclohexane, bp 73-74° (0.90 mm), n^{25}D 1.5653, is prepared in 83% yield by the addition of iodine azide to cyclohexene.[2] The ready availability of the requisite 2-iodoalkyl azides with defined stereochemistry as a consequence of the developments by Hassner and his co-workers[2] greatly enhances the value of the present synthesis of aziridines.
3. The ring closure of the intermediate β-iodoalkylamine can be effected by treating the amine in benzene with any one of several convenient bases. Excess anhydrous potassium carbonate is useful for cases where substituents are sensitive to aqueous base. Another alternative is treatment of the intermediate with one equivalent of *n*-butyllithium at room temperature. Because of the instability of *N*-phenylaziridines to aqueous base and/or heat, this is the preferred procedure for these derivatives.
4. The power of this synthesis is indicated by the stereospecific conversions of *cis*-2-butene to *N*-phenyl-*cis*-2,3-dimethylaziridine and *trans*-2-butene to *N*-phenyl-*trans*-2,3-dimethylaziridine.[1]
5. Aziridines are highly active physiologically and should be handled with great care.

[1] A. B. Levy and H. C. Brown, *J. Amer. Chem. Soc.*, **95**, 4067 (1973).
[2] F. W. Fowler, A. Hassner, and L. A. Levy, *J. Amer. Chem. Soc.*, **89**, 2077 (1967).

6.17. *CIS*-MYRTANYLMERCURIC CHLORIDE

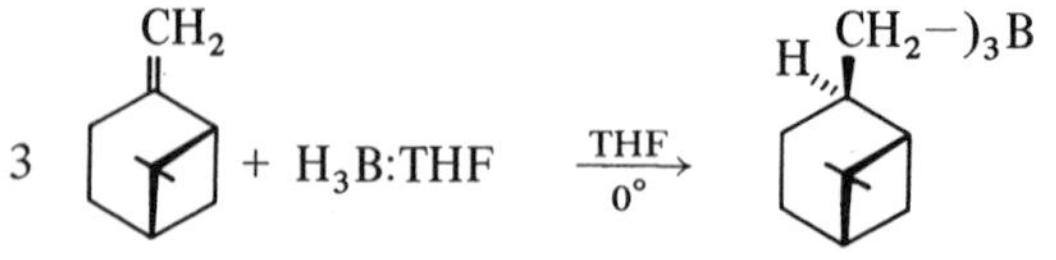

CH_2 H $CH_2-)_3B$

3 + H_3B:THF $\xrightarrow[0°]{THF}$

H $(CH_2-)_3B$ + 3 $Hg(OAc)_2$ $\xrightarrow[30\ min]{THF,\ 25°}$ 3 H CH_2HgOAc + $B(OAc)_3$

H CH_2HgOAc + NaCl ⟶ H CH_2HgCl + NaOAc

Procedure by Richard C. Larock[1]

Procedure

A dry 300-ml flask equipped with a septum inlet, thermometer well, side flask, magnetic stirring bar, and bubbler is assembled (**S9.1**). The side flask is attached by a short length of large-diameter flexible rubber tubing to permit addition of a solid (**S9.5**). In this flask is placed 31.9 g of mercuric acetate (0.100 mole). The assembly is flushed with nitrogen and then maintained under a static pressure. The flask is charged with 80 ml of tetrahydrofuran (**S9.11**) and 13.6 g (0.100 mole) of β-pinene (**P6.4:N1**). The flask is immersed in an ice bath, and hydroboration of the β-pinene is accomplished by the dropwise addition of 16.7 ml of a 2.00*M* solution of borane in tetrahydrofuran (**P2.3**) (**P2.4:N3**) over 0.5 hour to the rapidly stirred solution. After 1 hour at 25°, the mercuric acetate is added from the attached side flask to the stirred reaction mixture. After stirring for 30 minutes (**N1**), the suspension is poured into 400 ml of ice water, and 125 ml of a 1.0*M* solution of sodium chloride is added dropwise (**N2**). The solid is collected by filtration, washed with generous quantities of water, and dried overnight in a vacuum dessicator. There is obtained 36.5 g (98% yield) of *cis*-myrtanylmercuric chloride, mp 127-127.4° (**N3, N4**).

Notes

1. The time required for the conversion of the organoborane to the mercurial varies considerably with the structure. Thus the reaction with tri-*n*-hexylborane is complete in 1 minute, whereas *tris*-(3,3-dimethyl-1-butyl)borane requires 4 hours. The reactions of secondary alkyl derivatives are extraordinarily slow.[2]
2. Contact with the alkylmercuric acetates must be avoided as these compounds can cause severe blistering of the skin.
3. These derivatives are readily converted into the corresponding halides by treatment with bromine or iodine.[1]
4. The reaction is broadly applicable to terminal olefins which undergo

hydroboration. Thus methyl 10-undecenoate is converted into methyl 11-chloromercuriundecanoate in a yield of 91%. (The secondary alkyl isomeric moieties formed in the hydroboration do not react.) Dicyclohexylborane can be utilized to achieve selective hydroboration and the borane then converted into the mercurial. In this way, 4-vinylcyclohexene can be transformed into 2-(4-cyclohexenyl)ethylmercuric chloride in 93% yield.[1]

[1] R. C. Larock and H. C. Brown, *J. Amer. Chem. Soc.*, **92**, 2467 (1970).
[2] R. C. Larock and H. C. Brown, *J. Organometal. Chem.*, **26**, 35 (1971).

6.18. *TRANS*-CYCLOHEXYLETHENYLMERCURIC CHLORIDE

H C=C H B O O + $Hg(OAc)_2$ $\xrightarrow[0°]{THF}$

H C=C H HgOAc + O BOAc O

H C=C H HgOAc + NaCl $\xrightarrow{H_2O}$ H C=C H HgCl + NaOAc

Procedure by Richard C. Larock and Shyam K. Gupta[1]

Procedure

The apparatus consisting of a 100-ml flask equipped with a septum inlet, magnetic stirring bar, side flask [for introduction of a solid **(S9.5)**] and a mercury bubbler is assembled **(S9.1)**. In the side flask is placed 7.97 g (25 mmoles) of mercuric acetate. The apparatus is flushed with nitrogen. Then through the septum inlet 25 ml of tetrahydrofuran is introduced into the flask, followed by 5.7 g (25 mmole) of 2-*trans*-(cyclohexylethenyl)-1,3,2-benzodioxaborole (**P4.11**). The solution is cooled to 0°C and vigorously stirred as the mercuric acetate is introduced. The solid disappears in several minutes. The flask is separated from the assembly, and its contents are poured into 100 ml of ice water containing 2.0 g (35 mmoles) of sodium chloride (**N1**). The tetrahydrofuran is removed under vacuum, and the resulting white solid is filtered, washed very thoroughly with water, and dried overnight under vacuum. There is obtained 8.57 g (99%) of *trans*-2-cyclohexylethenylmercuric chloride, mp 134-135° (from 95% ethanol) (**N2, N3**).

Notes

1. Contact with the alkylmercuric acetates must be avoided as these compounds can cause severe blistering of the skin.
2. The synthesis can be carried out by directly using the reaction product from catecholborane and the acetylene without isolation of the addition product (see **P6.9**). The yield of product is still above 90%.
3. A wide variety of acetylenes can be accommodated:[1,2]

$$c\text{-}C_6H_{11}C{\equiv}C{-}CH_3 \longrightarrow (c\text{-}C_6H_{11})HC{=}C(CH_3)B(O_2C_6H_4) \longrightarrow (c\text{-}C_6H_{11})HC{=}C(CH_3)HgOAc$$

$$n\text{-}C_5H_{11}CH(OH)C{\equiv}CH \longrightarrow n\text{-}C_5H_{11}CH(OSiMe_2t\text{-}Bu)CH{=}CHB(O_2C_6H_4) \longrightarrow n\text{-}C_5H_{11}CH(OSiMe_2t\text{-}Bu)CH{=}CHHgCl$$

[1] R. C. Larock, S. K. Gupta, and H. C. Brown, *J. Amer. Chem. Soc.*, **94**, 4371 (1972).
[2] R. Pappo and P. W. Collins, *Tetrahedron Lett.*, 2627 (1972).

7

CARBON BOND FORMATION VIA ORGANOBORANES: SURVEY

Perhaps the most remarkable development in the organoborane field has been the unexpected utility of these derivatives to achieve carbon-carbon bond formation. This has been an area of intense research activity in the past few years. It is possible to give here only a highly concise survey of these rich developments.[1,2]

7.1. THE COUPLING REACTION

Treatment of organoboranes at 0° with alkali and aqueous silver nitrate results in the precipitation of metallic silver and the formation of the coupled product.[3,4] It is probable that the reaction proceeds through the formation of the silver alkyl, as in the related reaction involving Grignard reagents. However, this question has not yet been explored.

In many cases the reaction can be carried out in the hydroboration medium. With more hindered derivatives, there are advantages in replacing the hydroboration solvent with methanol which apparently facilitates reaction of the base with the more hindered organoboranes.

The reaction appears to be widely applicable. Terminal olefins undergo coupling with yields in the range of 60-80%.[3]

[1] H. C. Brown, *Boranes in Organic Chemistry,* Cornell University Press, Ithaca, N.Y., 1972.
[2] G. M. L. Cragg, *Organoboranes in Organic Synthesis,* Marcel Dekker, New York, 1973.
[3] H. C. Brown, N. C. Hébert, and C. H. Snyder, *J. Amer. Chem. Soc.,* 83, 1001 (1961).

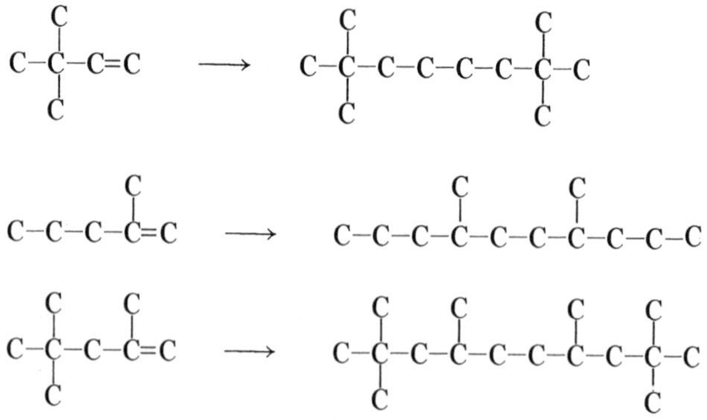

Internal olefins form the coupled products in lower yields, in the range of 35-50%.[4]

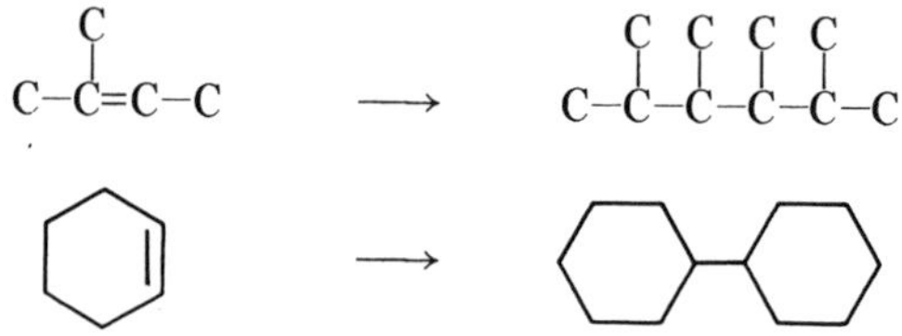

The reaction can be used without difficulties for a wide variety of structures.

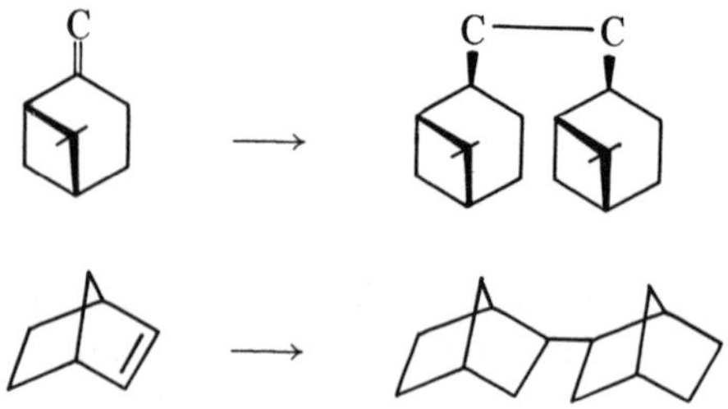

The coupling of two different alkyl groups is approximately statistical.[5] This means that the yields of products in the coupling of two such groups, R and R′, will be 25% R_2, 50% RR′, and 25% R'_2. However, the yield of the desired product, RR′, can be improved by using an excess of that olefin which is more readily available.

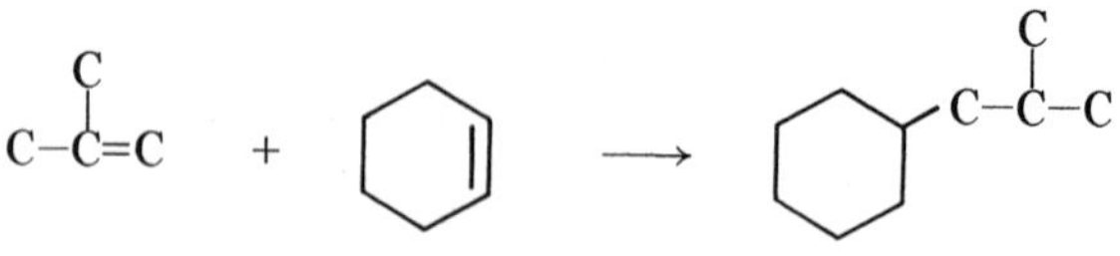

[4] H. C. Brown and C. H. Snyder, *J. Amer. Chem. Soc.*, 83, 1001 (1961).
[5] H. C. Brown, C. Verbrugge, and C. H. Snyder, *J. Amer. Chem. Soc.*, 83, 1002 (1961).

The reaction tolerates many functional groups not compatible with the Grignard reagent. This makes possible many simple one-step syntheses[6,7] (**P8.1**).

$$2\,CH_2{=}CH(CH_2)_8CO_2C_2H_5 \longrightarrow \begin{array}{l} CH_2CH_2(CH_2)_8CO_2C_2H_5 \\ | \\ CH_2CH_2(CH_2)_8CO_2C_2H_5 \end{array}$$

$$CH_3\overset{\displaystyle CH_3}{\overset{|}{C}}{=}CH_2 + CH_2{=}CH(CH_2)_8CO_2C_2H_5 \longrightarrow$$

$$(H_3C)_2CHCH_2CH_2CH_2(CH_2)_8CO_2C_2H_5$$

7.2. THE CYCLOPROPANE SYNTHESIS

Treatment of the γ-chloropropylboron moiety with base leads to the formation of cyclopropane.[8]

$$ClCH_2CH_2CH_2B\langle \xrightarrow{NaOH} \triangle + NaCl + HOB\langle$$

Unfortunately, the simple hydroboration of allyl chlorides results in a large fraction of the boron going to the nonallylic position[9] (**S3**).

$$\underbrace{\overset{\displaystyle CH_2{=}CH}{\uparrow\quad\uparrow}}_{\substack{50\%\quad 50\% \\ BH_3}}{-}CH_2Cl$$

9-BBN not only overcomes this unfavorable directive influence of the allylic chlorine substituent, but the steric availability of the boron atom apparently favors coordination with the base leading to the desired ring closure.[10]

$$CH_2{=}CH\underset{\displaystyle Cl}{\underset{|}{C}}H_2 \xrightarrow{\text{9-BBN (BH)}} (\text{9-BBN})B{-}CH_2CH_2\underset{\displaystyle Cl}{\underset{|}{C}}H_2 \xrightarrow{OH^-} \triangle$$

[6] C. Verbrugge, Ph.D. thesis, Purdue University.
[7] Research with M. K. Unni.
[8] M. F. Hawthorne and J. A. Dupont, *J. Amer. Chem. Soc.*, **80**, 5830 (1958).
[9] H. C. Brown and K. A. Keblys, *J. Amer. Chem. Soc.*, **86**, 1791 (1964).
[10] H. C. Brown and S. P. Rhodes, *J. Amer. Chem. Soc.*, **91**, 2149 (1969).

This synthesis of cyclopropane derivatives (**P8.2**) appears to be broadly applicable, as indicated by the following examples.

$$CH_2{=}CHC(CH_3)(Cl){-}CH_3 \longrightarrow \text{1,1-dimethylcyclopropane}$$

$$CH_2{=}CHC(H)(Cl){-}Cl \longrightarrow \text{chlorocyclopropane}$$

$$CH_2{=}CHCH(Cl)CH_2Cl \longrightarrow \text{cyclopropyl}{-}CH_2Cl$$

It is also possible to utilize this approach to synthesize B-cyclopropyl-9-BBN[11] (**P4.6**) as well as the related B-cyclobutyl-,[11] B-cyclopentyl-,[12] and B-cyclohexyl-9-BBN[12] structures.

$$HC{\equiv}CCH_2Br \xrightarrow{\text{9-BBN (BH)}} HC(B)_2CH_2CH_2Br \xrightarrow{MeLi} \text{9-BBN}{-}B{-}\text{cyclopropyl}$$

7.3. CARBONYLATION TO TERTIARY ALCOHOLS

It was originally noted that trialkylboranes react with carbon monoxide at high pressures to give products oxidizable to trialkylcarbinols.[13]

$$R_3B \xrightarrow[10{,}000\ psi]{CO} R_3CBO \xrightarrow[NaOH]{H_2O_2} R_3COH$$

Fortunately, in many cases the reaction can be carried out in glass at

[11] H. C. Brown and S. P. Rhodes, *J. Amer. Chem. Soc.*, **91**, 4306 (1969).
[12] C. G. Scouten, Ph.D. thesis, Purdue University.
[13] M. E. D. Hillman, *J. Amer. Chem. Soc.*, **84**, 4715 (1962).

atmospheric pressure,[14] especially conveniently utilizing the automatic hydrogenator[15] modified for carbonylation[16] **(S9.4)**. It has proved possible to control the reaction to provide a highly useful synthetic route not only to tertiary alcohols of a wide variety of structures, but also to ketones, secondary alcohols, carboxylic acids, methylol derivatives, and aldehydes. It is possible here to survey only some of the highlights. For a more complete exposition, other reviews should be consulted.[1,17]

At a temperature of 100-125° many organoboranes absorb 1 mole of carbon monoxide at atmospheric pressure, yielding an intermediate which can be oxidized to the tertiary alcohol with alkaline hydrogen peroxide **(P8.3)**. The reaction appears to be broadly applicable.[14]

$$CH_3CH{=}CHCH_3 \longrightarrow (CH_3CH_2\overset{CH_3}{\overset{|}{C}H})_3COH \qquad 87\%$$

$$\text{cyclopentene} \longrightarrow (\text{cyclopentyl})_3COH \qquad 90\%$$

$$\text{cyclohexene} \longrightarrow (\text{cyclohexyl})_3COH \qquad 80\%$$

$$\text{norbornene} \longrightarrow (\text{norbornyl})_3COH \qquad 80\%$$

This development makes readily available trialkylcarbinols containing bulky alkyl groups. For example, the yield of tricyclohexylcarbinol through the Grignard reaction is reported to be only 7%, raised to 19% by the use of a special procedure involving sodium.[18]

7.4. CARBONYLATION TO KETONES

The reaction appears to involve the following steps:[13]

$$R_3B + CO \rightleftharpoons R_3\overset{-}{B}{:}\overset{+}{C}O$$

[14] H. C. Brown and M. W. Rathke, *J. Amer. Chem. Soc.*, 89, 2737 (1967).
[15] C. A. Brown and H. C. Brown, *J. Amer. Chem. Soc.*, 84, 2829 (1962).
[16] M. W. Rathke and H. C. Brown, *J. Amer. Chem. Soc.*, 88, 2606 (1966).
[17] H. C. Brown, *Acc. Chem. Res.*, 2, 65 (1969).
[18] P. D. Bartlett and A. Schneider, *J. Amer. Chem. Soc.*, 67, 141 (1945).

$$R_3\overset{-}{B}{:}\overset{+}{C}O \rightarrow R_2B{-}\underset{\underset{O}{\|}}{C}{-}R$$

$$R_2B\underset{\underset{O}{\|}}{C}R \longrightarrow RB\!\!-\!\!\!\!\!\underset{O}{\smile}\!\!\!\!\!-\!\!CR_2$$

$$RB\!\!-\!\!\!\!\!\underset{O}{\smile}\!\!\!\!\!-\!\!CR_2 \longrightarrow OBCR_3$$

The presence of water in the reaction mixture (it should be recalled that trialkylboranes are stable to water) makes it possible to halt the reaction prior to the migration of the third alkyl group. Apparently the water converts the boraepoxide to a boraglycol. Oxidation of the intermediate provides the corresponding ketone; hydrolysis with alkali yields the secondary carbinol.

$$R_3B + CO \longrightarrow RB\!\!-\!\!\!\!\!\underset{O}{\smile}\!\!\!\!\!-\!\!CR_2 \xrightarrow{H_2O} \underset{OH}{RB}{-}\underset{OH}{CR_2} \begin{array}{l} \xrightarrow{-OH} R_2CHOH \\ \xrightarrow{[O]} R_2CO \end{array}$$

The procedure provides a convenient synthesis of either secondary carbinols or ketones.[19]

cyclopentene $\longrightarrow$ (cyclopentyl$)_2CO$ 90%

norbornene $\longrightarrow$ (norbornyl$)_2CO$ 82%

The reaction involves the loss of one of the three R groups. Fortunately, the thexyl group exhibits a reluctance to migrate from boron to carbon. Thus the thexyl derivatives can be utilized for ketone synthesis, taking advantage of the unique hydroborating characteristics of thexylborane[20] (**S3.6**).

$$\text{thexyl}{-}BH_2 \xrightarrow{(CH_3)_2C{=}CH_2} \text{thexyl}{-}B\begin{matrix} H \\ CH_2CH(CH_3)_2 \end{matrix} \xrightarrow{RCH{=}CH_2}$$

[19] H. C. Brown and M. W. Rathke, *J. Amer. Chem. Soc.*, 89, 2738 (1967).
[20] H. C. Brown and E. Negishi, *J. Amer. Chem. Soc.*, 89, 5285 (1967).

$$\text{thexyl-B}(CH_2CH_2R)(CH_2CH(CH_3)_2) \xrightarrow[H_2O]{CO} \xrightarrow{[O]} O{=}C(CH_2CH_2R)(CH_2CH(CH_3)_2)$$

Many functional groups can be tolerated in this ketone synthesis.

$$(CH_3)_2C{=}CH_2 + CH_2{=}CHCH_2CO_2R \longrightarrow (CH_3)_2CHCH_2\overset{O}{\overset{\|}{C}}CH_2CH_2CH_2CO_2R$$

84%

$$\text{norbornene} + CH_2{=}CHCH_2Cl \longrightarrow \text{2-norbornyl}{-}\overset{O}{\overset{\|}{C}}{-}CH_2CH_2CH_2Cl$$

63%

Cyclic ketones are readily synthesized from dienes.[21]

$$\text{1,4-pentadiene (2-methyl)} \xrightarrow{\text{thexyl-}BH_2} \text{B-thexylborolane} \longrightarrow \text{2-methylcyclopentanone (C=O)}$$

75%

$$\text{1,6-heptadiene} \xrightarrow{\text{thexyl-}BH_2} \text{B-thexylborepane} \longrightarrow \text{cycloheptanone (C=O)}$$

65%

This development makes possible a new annelation reaction, as shown for the stereospecific conversion of cyclohexanone into the thermodynamically disfavored *trans*-perhydro-1-indanone[22] **(P8.6)**.

$$\text{cyclohexanone} \xrightarrow[2.\ H_2O]{1.\ CH_2{=}CHMgX} \text{1-vinylcyclohexanol (HO)} \xrightarrow{-H_2O} \text{1-vinylcyclohexene}$$

$$\xrightarrow{\text{thexyl-}BH_2} \text{trans-B-thexyl-perhydroboraindane (H, H)} \xrightarrow[H_2O]{CO} \xrightarrow{[O]} \textit{trans}\text{-perhydro-1-indanone (H, H, O)}$$

[21] H. C. Brown and E. Negishi, *J. Amer. Chem. Soc.*, 89, 5477 (1967).
[22] H. C. Brown and E. Negishi, *Chem. Comm.*, 594 (1968).

The reaction appears to be relatively broadly applicable, as indicated by the structures which have been synthesized.

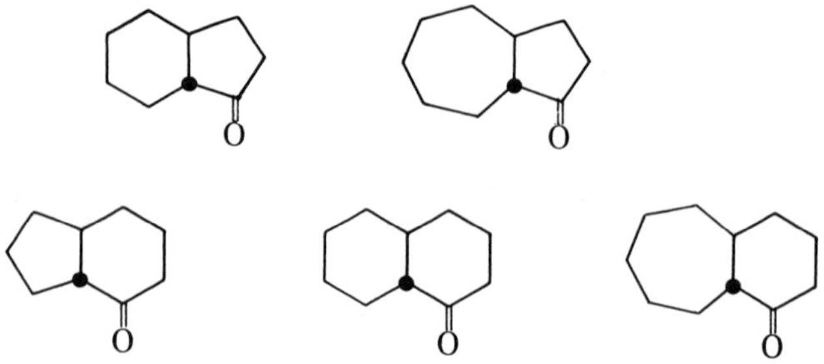

Unfortunately, the rate of reaction of these thexyl derivatives with carbon monoxide at atmospheric pressure is relatively low. Moreover, it is undesirable to subject such labile structures to elevated temperatures for a relatively long reaction time. Consequently, it is recommended that these derivatives be carbonylated at 1000 psi (**P8.4**). Fortunately, it is possible to convert these thexyl derivatives to the corresponding ketones under milder conditions using the Pelter cyanidation reaction (**S7.7**, **P8.8**).

7.5. CARBONYLATION TO ALDEHYDES

It is possible to halt the reaction of organoboranes with carbon monoxide after the transfer of but one alkyl group, provided that an active hydride reagent, such as lithium trimethoxyaluminohydride, is present.[23] Hydrolysis of the intermediate yields the methylol derivative; oxidation yields the aldehyde.

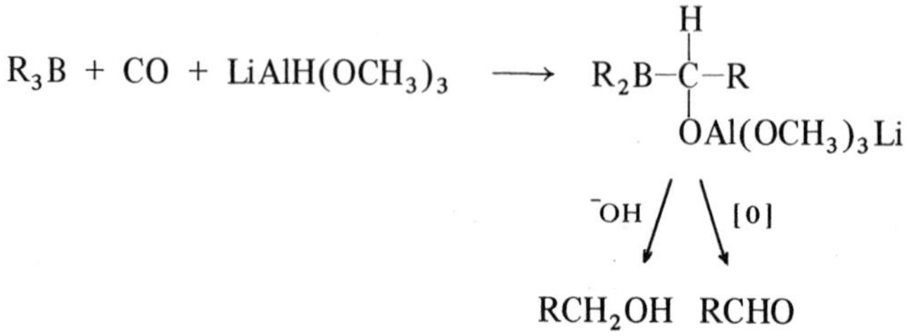

$$R_3B + CO + LiAlH(OCH_3)_3 \longrightarrow R_2B{-}\underset{\displaystyle OAl(OCH_3)_3Li}{\overset{\displaystyle H}{C}}{-}R$$

$$\xrightarrow{^{-}OH} RCH_2OH \qquad \xrightarrow{[O]} RCHO$$

The loss of the two R groups can be avoided by employing 9-BBN[24] (**P8.5**).

[23] H. C. Brown, R. A. Coleman, and M. W. Rathke, *J. Amer. Chem. Soc.*, **90**, 499 (1968).
[24] H. C. Brown, E. F. Knights, and R. A. Coleman, *J. Amer. Chem. Soc.*, **91**, 2144 (1969).

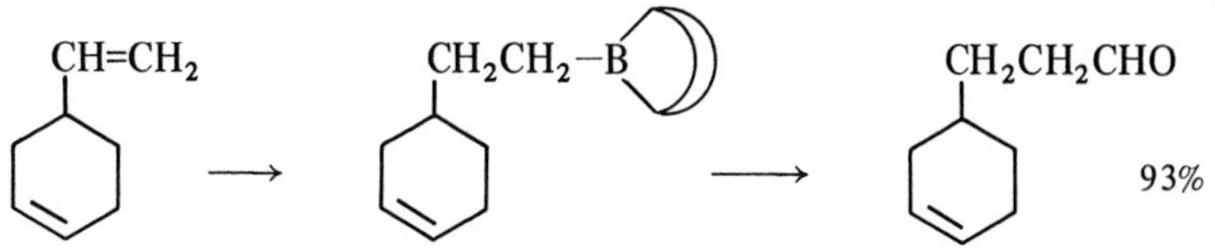

$\xrightarrow[\text{CO, 0°}]{\text{LiAl(MeO)}_3\text{H}}$ $\xrightarrow{[O]}$ CHO

79%

$CH=CH_2$ → CH_2CH_2-B → CH_2CH_2CHO

93%

The reaction can also be adapted to introduce the aldehyde group into molecules containing functional substituents[25] (**P8.5**).

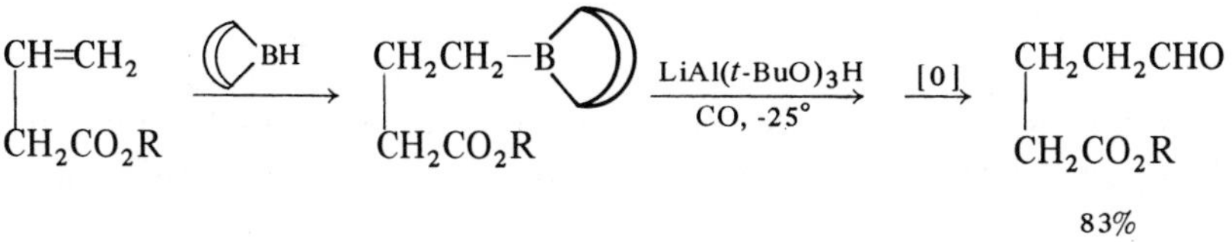

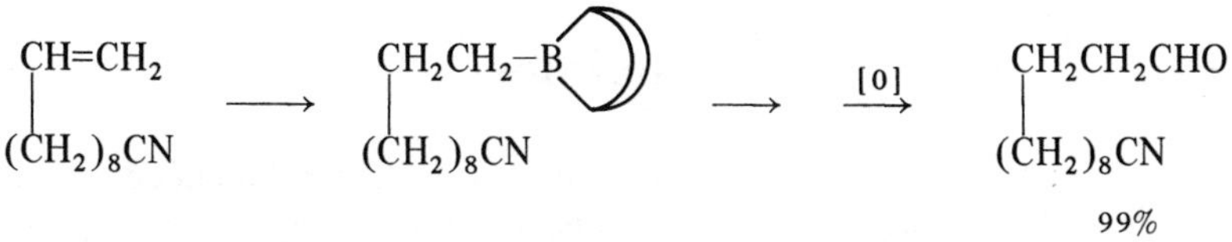

$CH=CH_2$ / CH_2CO_2R → CH_2CH_2-B / CH_2CO_2R $\xrightarrow[\text{CO, -25°}]{\text{LiAl}(t\text{-BuO})_3\text{H}}$ $\xrightarrow{[O]}$ CH_2CH_2CHO / CH_2CO_2R

83%

$CH=CH_2$ / $(CH_2)_8CN$ → CH_2CH_2-B / $(CH_2)_8CN$ → $\xrightarrow{[O]}$ CH_2CH_2CHO / $(CH_2)_8CN$

99%

7.6. STITCHING AND RIVETING

The remarkable ease with which boron can be replaced by carbon in these reactions makes possible new simple syntheses of polycyclic structures[26-28] (**P8.7**).

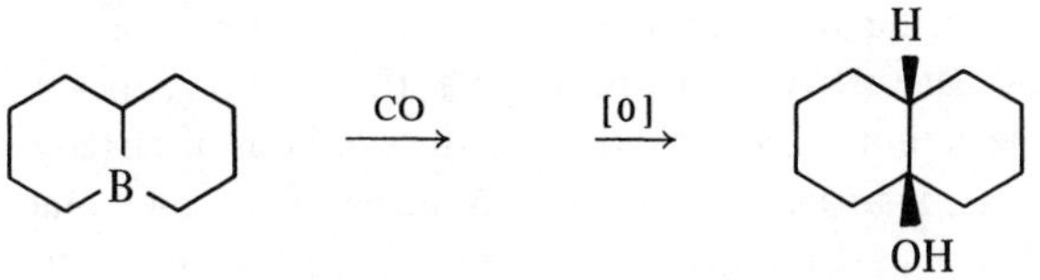

$\xrightarrow{\text{CO}}$ $\xrightarrow{[O]}$ H / OH

Predominantly *cis*

[25]H. C. Brown and R. A. Coleman, *J. Amer. Chem. Soc.*, **91**, 4606 (1969).
[26]H. C. Brown and E. Negishi, *J. Amer. Chem. Soc.*, **91**, 1224 (1969).
[27]E. F. Knights and H. C. Brown, *J. Amer. Chem. Soc.*, **90**, 5283 (1968).
[28]H. C. Brown and E. Negishi, *J. Amer. Chem. Soc.*, **89**, 5478 (1967).

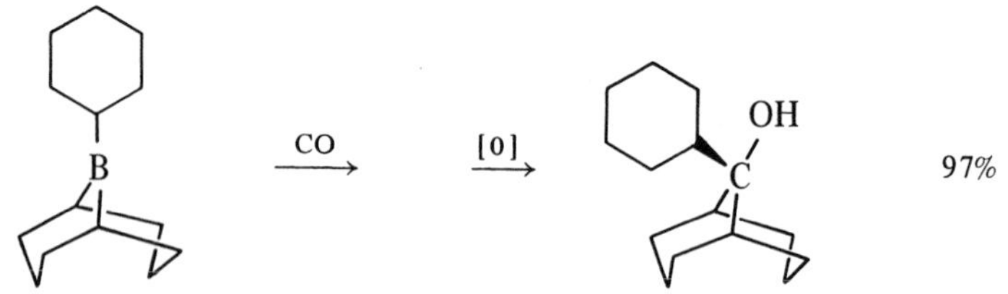

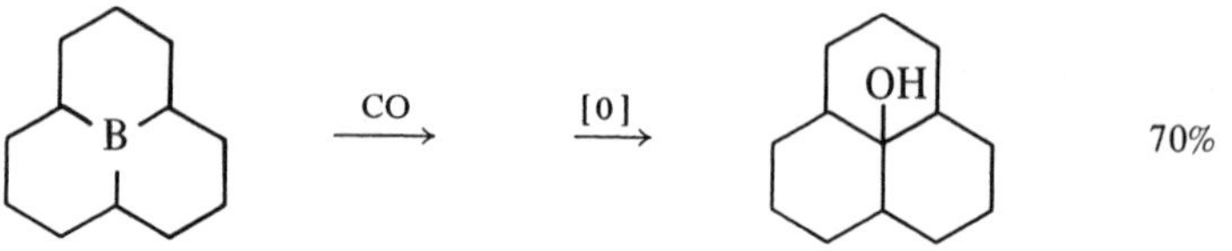

$\xrightarrow{CO}$ $\xrightarrow{[O]}$ 97%

$\xrightarrow{CO}$ $\xrightarrow{[O]}$ 70%

Thus it is now possible to take relatively open, readily available structures, "stitch" them together with boron, and then "rivet" them with carbon into more complex, difficultly accessible structures.

7.7. CYANIDATION

The cyanide ion is isoelectronic with carbon monoxide. Organoboranes do react with alkali metal cyanides, but the products are the simple coordination compounds. Treatment of these addition compounds with a suitable electrophilic reagent, such as trifluoroacetic anhydride, brings about migration of groups from boron to carbon.[29]

$$[n\text{-}Bu_3BCN]^-Na^+ \xrightarrow[\text{-78°, then 45° for 12 hr}]{(CF_3CO)_2O} [\ ] \xrightarrow{[O]} n\text{-}Bu_3COH$$

At present, the transfer of three alkyl groups from boron appears to occur readily only for primary alkyls.[30] In this respect, the reaction appears to be less convenient than the carbonylation reaction for the synthesis of tertiary alcohols.

On the other hand, it is possible to take advantage of the preference for the migration of two groups to provide the corresponding ketones. Indeed, the thexyldialkylboranes react readily, providing a highly convenient route to such ketones[31] (**P8.8**).

[29] A. Pelter, M. G. Hutchings, and K. Smith, *Chem. Commun.*, 1529 (1970).
[30] A. Pelter, M. G. Hutchings, and K. Smith, *Chem. Commun.*, 1048 (1971).
[31] A. Pelter, M. G. Hutchings, and K. Smith, *Chem. Commun.*, 1048 (1971).

$$\left[\text{thexyl}-\underset{R}{\overset{R'}{B}}-CN \right]^- Na^+ \xrightarrow{(CF_3CO)_2O} [\ \ ] \xrightarrow{[O]} RCOR'$$

It should be recalled that these thexyl derivatives are best carbonylated with carbon monoxide at elevated pressure (1000 psi). Consequently, the cyanidation reaction offers major advantages in this application.

7.8. DCME REACTION

Organoboranes also undergo a facile reaction with haloforms in the presence of strong, sterically demanding bases.[32]

$$R_3B + HCCl_3 + 2\,LiOCEt_3 \xrightarrow[67°,\ 1\ hr]{THF} R_3CB\begin{matrix} Cl \\ OCEt_3 \end{matrix} + LiCl + Et_3COH$$

Unfortunately, the intermediates produced, while readily oxidized for primary alkyl groups, proved resistant for R being isoalkyl or *sec*-alkyl groups.[33]

Fortunately, the corresponding reaction with α,α-dichloromethyl methyl ether is far more favorable—it requires only one equivalent of base, proceeds in only a few minutes at 25° or even 0°. and the intermediate produced is readily oxidized.[34]

$$R_3B + HCCl_2OCH_3 + LiOCEt_3 \xrightarrow[25°,\ 15\ min]{THF} R_3CB\begin{matrix} Cl \\ OCH_3 \end{matrix} + LiCl + Et_3COH$$

No difficulty was encountered in applying the reaction to a wide variety of organoboranes.[34]

$$(\text{cyclopentyl})_3B \longrightarrow \longrightarrow (\text{cyclopentyl})_3COH \qquad 97\%$$

[32] H. C. Brown, B. A. Carlson, and R. H. Prager, *J. Amer. Chem. Soc.*, **93**, 2070 (1971).
[33] H. C. Brown and B. A. Carlson, *J. Organometal. Chem.*, **54**, 61 (1973).
[34] H. C. Brown and B. A. Carlson, *J. Org. Chem.*, **38**, 2422 (1973).

95%

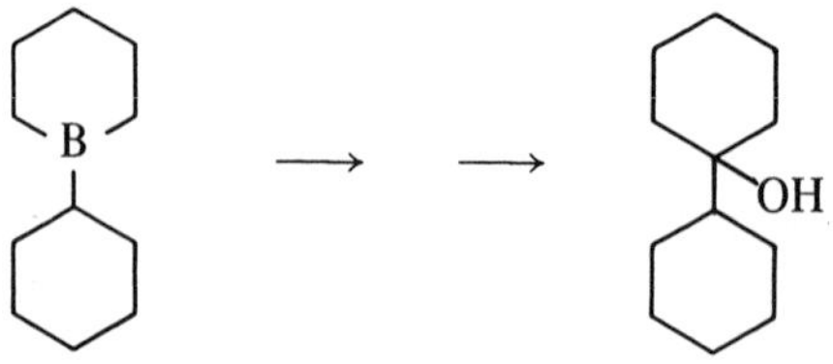

71%

Indeed, it has proved capable of handling exceptional hindered systems[35] (**P8.9**).

Dialkylborinic acid esters, now readily available via hydroboration with chloroborane-ethyl etherate (**S3.5**), react readily with the reagent[36] (**P8.10**).

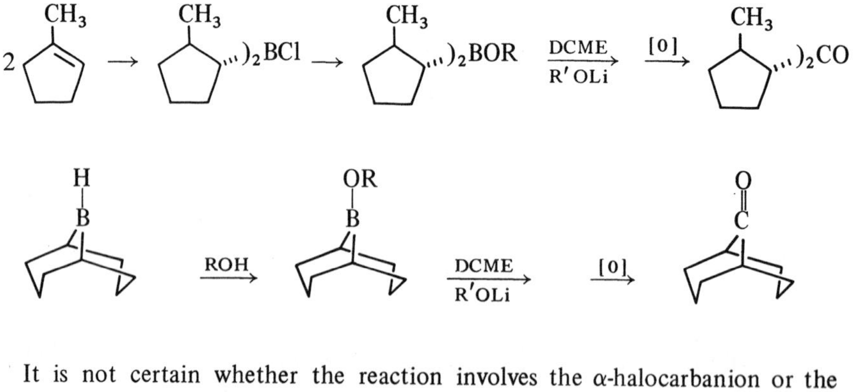

It is not certain whether the reaction involves the α-halocarbanion or the corresponding carbene. Assuming the carbanion, the reaction would involve the following transformations:

$$R_3B \xrightarrow{Li^{+}\,{}^{-}CCl_2OCH_3} [R_2\overset{R}{\overset{|}{B}}-CCl_2OCH_3]^{-}Li^{+} \xrightarrow{-LiCl} R_2B-\overset{R}{\overset{|}{C}}Cl(OCH_3)$$

$$R-\underset{R}{\underset{|}{B}}-\overset{R}{\overset{|}{C}}Cl(OCH_3) \xrightarrow{THF} R-\overset{OCH_3}{\overset{|}{B}}-CR_2Cl \xrightarrow{THF} Cl(OCH_3)BCR_3 \xrightarrow{[O]} HOCR_3$$

[35]H. C. Brown, J.-J. Katz, and B. A. Carlson, *J. Org. Chem.*, **38**, 3968 (1973).
[36]B. A. Carlson and H. C. Brown, *J. Amer. Chem. Soc.*, **95**, 6876 (1973).

On this basis, the borinic ester reaction ($R = OCH_3$) would proceed:

$$R_2BOCH_3 \xrightarrow{CCl_2OCH_3} [R_2(CH_3O)B{-}CCl_2OCH_3]^-Li^+ \xrightarrow{-LiCl}$$

$$R(CH_3O)B{-}CRCl(OCH_3) \xrightarrow{THF} (CH_3O)_2B{-}CR_2Cl$$

Oxidation of the latter species produces the ketone. It has proved possible to isolate and characterize these α-chloroboronic acid esters.[37]

7.9. α-ALKYLATION AND ARYLATION

In the past the customary procedure for the introduction of a carbon group in the α-position of esters, ketones, nitriles, and similar activated derivatives involved formation of the carbanion by removal of an activated α-hydrogen atom, followed by reaction of the carbanion with a suitable organic halide. Only groups which are capable of undergoing facile SN_2 displacements can be utilized. Groups, such as 2-norbornyl or phenyl, cannot be introduced in this manner. Fortunately, this restriction can be circumvented by use of the organoboranes in their base induced reaction with α-halo esters, ketones, nitriles, and similar derivatives. Again, the present discussion must be very brief. More complete surveys are available.[1,2,38]

Treatment of an α-halo ester, ketone, nitrile, and related derivatives, with a suitable base in the presence of an organoborane results in the replacement of the halogen atom by an alkyl or aryl group of the organoborane.[39,40]

$$R_3B + CH_2BrCO_2C_2H_5 + t\text{-BuOK} \longrightarrow RCH_2CO_2C_2H_5 + t\text{-BuOBR}_2 + KBr$$

The halogens of α,α-dihalo derivatives can be replaced consecutively.[41]

$$CHBr_2CO_2C_2H_5 \longrightarrow RCHBrCO_2C_2H_5 \longrightarrow RR'CHCO_2C_2H_5$$

Consequently, this reaction provides a convenient route to α-halo esters and an alternative to the malonic ester route to dialkylacetic acids.

The reaction appears to involve conversion of the α-halo derivative into the

[37] B. A. Carlson, J.-J. Katz, and H. C. Brown, *J. Organomet. Chem.*, **67**, C39 (1974).
[38] H. C. Brown and M. M. Rogić, *Organomet. Chem. Syn.*, **1**, 305 (1972).
[39] H. C. Brown, M. M. Rogić, M. W. Rathke, and G. W. Kabalka, *J. Amer. Chem. Soc.*, **90**, 818 (1968).
[40] H. C. Brown, M. M. Rogić, and M. W. Rathke, *J. Amer. Chem. Soc.*, **90**, 6218 (1968).
[41] H. C. Brown, M. M. Rogić, M. W. Rathke, and G. W. Kabalka, *J. Amer. Chem. Soc.*, **90**, 1911 (1968).

corresponding anion, formation of the boron "ate" complex, transfer of the organic group from boron to carbon, and protonolysis of the intermediate.

$$t\text{-BuO}^-K^+ + CH_2BrCO_2C_2H_5 \longrightarrow K^+[^-CHBrCO_2C_2H_5] + t\text{-BuOH}$$

$$R_3B + K^+[^-CHBrCO_2C_2H_5] \longrightarrow K^+[R_3BCHBrCO_2C_2H_5]^-$$

$$K^+\left[R{-}\overset{R}{\underset{R}{B^-}}{-}\underset{Br}{C}HCO_2C_2H_5\right] \longrightarrow R_2B{-}\overset{R}{C}HCO_2C_2H_5 + KBr$$

$$R_2B{-}\overset{R}{C}HCO_2C_2H_5 \longrightarrow RCH{=}\overset{OBR_2}{C}{-}OC_2H_5$$

$$RCH{=}\overset{OBR_2}{C}{-}OC_2H_5 + t\text{-BuOH} \longrightarrow RCH_2CO_2C_2H_5 + t\text{-BuOBR}_2$$

A major disadvantage is the utilization of only one of the three groups of the organoborane. Fortunately, this difficulty is overcome by the use of the B-organo-9-BBN derivatives **(P4.4, P4.7) (P8.11)**.

$$C_6H_5COCH_2Br + H_3C{-}C_6H_4{-}B(9\text{-BBN}) + t\text{-BuOK} \xrightarrow{0°} C_6H_5COCH_2{-}C_6H_4{-}CH_3 + t\text{-BuO}{-}B(9\text{-BBN}) + KBr$$

95%

Furthermore, the replacement of potassium t-butoxide by the milder, more sterically demanding base, potassium 2,6-di-t-butylphenoxide, has major advantages **(P8.12)**. These developments make possible the introduction of a wide variety of alkyl and aryl groups into the α-position of a wide variety of derivatives.

Among the groups which are readily transferred from the B-R-9-BBN derivatives are methyl, cyclohexyl, *exo*-norbornyl, phenyl, and p-tolyl. Among the groups which undergo replacement of the halogen are α-haloketones,[42]

[42] H. C. Brown, H. Nambu, and M. M. Rogić, *J. Amer. Chem. Soc.*, **91**, 6852 (1969).

α-halo esters,[44] α-halo nitriles,[43] α-dihalo esters,[44] α-dihalo nitriles,[45] and halomalononitriles.[46]

$CHBr_2CO_2C_2H_5$ → (B-cyclopentyl-9-BBN) → cyclopentyl-$CHBrCO_2C_2H_5$ 78%

$CH_2BrCOCH_3$ → (B-phenyl-9-BBN) → phenyl-CH_2COCH_3 76%

CH_2ClCN → (B-norbornyl-9-BBN) → norbornyl-CH_2CN 65%

$CHCl_2CN$ → (B-cyclohexyl-9-BBN) → (B-*sec*-Bu-9-BBN) → $C_2H_5CH(CH_3)$–CH(cyclohexyl)–CN 61%

$BrCH(CN)_2$ → (B-cyclopentyl-9-BBN) → cyclopentyl-$CH(CN)_2$ 87%

It should be noted that the next to last example involves the successive introduction of two secondary alkyl groups, a reaction that fails in the usual malonic ester synthesis.

Finally, extension of the reaction to 4-bromocrotonic acid provides a convenient procedure for a four carbon atom homologation[47] **(P8.12)**.

7.10. α-BROMINATION TRANSFER

The above results suggest that the migration of groups from boron to carbon in α-haloorganoboranes is a facile reaction which occurs under very mild conditions.

[43] H. C. Brown, H. Nambu, and M. M. Rogić, *J. Amer. Chem. Soc.*, **91**, 6854 (1969).
[44] H. C. Brown, H. Nambu, and M. M. Rogić, *J. Amer. Chem. Soc.*, **91**, 6855 (1969).
[45] H. Nambu and H. C. Brown, *J. Amer. Chem. Soc.*, **92**, 5790 (1970).
[46] H. Nambu and H. C. Brown, *Organomet. Chem. Syn.*, **1**, 95 (1970).
[47] H. C. Brown and H. Nambu, *J. Amer. Chem. Soc.*, **92**, 1761 (1970).

$$-\overset{|}{\underset{X}{C}}-\overset{R}{\overset{|}{B}}- \xrightarrow[0°]{\text{base}} -\overset{R}{\overset{|}{\underset{|}{C}}}-\underset{\text{base}}{\underset{|}{B}}- + X^-$$

In many cases, even tetrahydrofuran is adequate to serve as the base. This development provides a promising new route to the synthesis of carbon structures.

One approach to the α-bromo intermediate is the reaction of organoboranes with α-halocarbanions, reviewed above. A second approach is the hydroboration of appropriate vinyl halides (**S1.7**). A third approach which has proved unexpectedly versatile is the photochemical bromination of organoboranes.

$$R'-\overset{|}{\underset{H}{C}}-\overset{R}{\overset{|}{B}}- \xrightarrow[h\nu]{Br_2} R'-\overset{|}{\underset{Br}{C}}-\overset{R}{\overset{|}{B}}- \xrightarrow{\text{base}} R'-\overset{R}{\overset{|}{\underset{|}{C}}}-\underset{\text{base}}{\underset{|}{B}}-$$

Fortunately, the carbon-boron bond of trialkylboranes is remarkably stable to the action of bromine (**S5.6**), and the α-position is remarkably active toward attack by bromine atoms. For example, in a comparison the α-position of B-isopropyl-9-BBN is 5.5 times as reactive as the α-position in cumene and some 600 times as reactive as the tertiary position in isobutane.[48]

B-isopropyl-9-BBN: $B-\overset{CH_3}{\overset{|}{\underset{H}{C}}}-CH_3$	Cumene: $C_6H_5-\overset{CH_3}{\overset{|}{\underset{H}{C}}}-CH_3$	$H_3C-\overset{CH_3}{\overset{|}{\underset{H}{C}}}-CH_3$
↑	↑	↑
600	120	1.00

Bromination of the organoborane in the presence of water results in an immediate migration of the initially formed derivative.[49] Thus the photochemical bromination of tri-*sec*-butylborane in the presence of water with one molar equivalent of bromine results in the union of two secondary butyl groups.

$$sec\text{-}Bu_3B \xrightarrow[h\nu]{Br_2} sec\text{-}Bu_2B\overset{CH_3}{\overset{|}{\underset{Br}{\underset{|}{C}}}}CH_2CH_3 \longrightarrow \begin{array}{c} CH_3CH_2CHCH_3 \\ | \\ CH_3CH_2CCH_3 \\ | \\ B \\ sec\text{-}Bu \diagup \quad \diagdown OH \end{array}$$

[48]H. C. Brown and N. R. De Lue, *J. Amer. Chem. Soc.*, **96**, 311 (1974).
[49]C. F. Lane and H. C. Brown, *J. Amer. Chem. Soc.*, **93**, 1025 (1971).

Oxidation produces an 86% yield of 3,4-dimethyl-3-hexanol. Further bromination results in the migration of the third *sec*-butyl group.

$$\xrightarrow[H_2O]{Br_2,\,h\nu} \begin{array}{l} CH_3CH_2CHCH_3 \\ CH_3CH_2CCH_3 \\ CH_3CH_2CCH_3 \\ \qquad\quad B(OH)_2 \end{array} \xrightarrow{[O]} \begin{array}{l} CH_3CH_2CHCH_3 \\ CH_3CH_2CCH_3 \\ CH_3CH_2CCH_3 \\ \qquad\quad OH \end{array}$$

The reaction takes an alternate pathway in the case of organoboranes having primary alkyl groups, such as triethylborane and tri-*n*-butylborane. In the case of these derivatives, following the first migration, substitution occurs at the tertiary position of the newly formed group rather than at the secondary position of the remaining unmigrated group.

$$(CH_3CH_2)_2B \xrightarrow[h\nu]{Br_2} (CH_3CH_2)_2B\underset{Br}{C}HCH_3 \xrightarrow{H_2O} CH_3CH_2B\overset{CH_2CH_3}{\underset{OH}{C}}HCH_3$$

$$\xrightarrow[h\nu]{Br_2} CH_3CH_2\underset{HO}{B}\overset{CH_2CH_3}{\underset{Br}{C}}CH_3 \xrightarrow{H_2O} (HO)_2B-\overset{CH_2CH_3}{\underset{CH_2CH_3}{C}}CH_3$$

It is now feasible to hydroborate with monochloroborane and synthesize R_2BCl **(S3.5)**. These derivatives are useful intermediates in this synthetic procedure. For example, bromination-hydrolysis of dicyclopentylboron chloride **(P4.14)** proceeds to yield (97%) 1-cyclopentylcyclopentanol.[50]

$$(C_5H_9)_2BCl \xrightarrow{H_2O} (C_5H_9)_2BOH \xrightarrow[h\nu]{Br_2} C_5H_8(Br)-B(OH)-C_5H_9$$

$$\xrightarrow{H_2O} (C_5H_9)C_5H_8-B(OH)_2 \xrightarrow{[O]} (C_5H_9)C_5H_8-OH$$

[50] H. C. Brown, Y. Yamamoto, and C. F. Lane, *Synthesis*, 303 (1972).

The thexyl group in thexyldialkylboranes does not contain any hydrogen atom *alpha* to the boron. Moreover, it exhibits the same reluctance to migrate that it exhibits in the carbonylation and cyanidation reactions.[20,31] Consequently, one can utilize the unique hydroboration characteristics of thexylborane **(S3.6)** and then use bromination-hydrolysis to link two groups together[51] **(P8.13)**.

$CH=CH_2$ / CH_2CO_2R; $(CH_2)_3CO_2R$; $Br_2, h\nu$ / H_2O; OH; $(CH_2)_3CO_2R$; [O]; HO $(CH_2)_3CO_2R$; 90%

For optimum results, either with the dialkylborinic acids[50] or the thexyldialkylboranes[51] at least one of the groups should have a tertiary hydrogen atom alpha to the boron to provide a point of selective attack.

Finally, it is possible to apply this synthesis to the many polycyclic organoborane structures which can now be readily synthesized[52,53] **(P8.14)**.

CH_3; B; H_2O / $Br_2, h\nu$; CH_3 OH; B; H; [O]; OH; H

7.11. REACTION WITH DIAZO COMPOUNDS

The reaction of organoboranes with diazo derivatives provides another valuable means for forming carbon-carbon bonds under mild conditions.[54-56] The

[51] H. C. Brown, Y. Yamamoto, and C. F. Lane, *Synthesis*, 304 (1972).
[52] Y. Yamamoto and H. C. Brown, *Chem. Commun.*, 801 (1973).
[53] Y. Yamamoto and H. C. Brown, *J. Org. Chem.*, **39**, 861 (1974).
[54] J. Hooz and S. Linke, *J. Amer. Chem. Soc.*, **90**, 5936 (1968).
[55] J. Hooz and S. Linke, *J. Amer. Chem. Soc.*, **90**, 6891 (1969).
[56] J. Hooz and G. F. Morrison, *Can. J. Chem.*, **48**, 868 (1970).

reaction has been demonstrated for diazoacetaldehyde,[56] diazoacetone,[54] diazoacetonitrile,[55] and ethyl diazoacetate.[55] Consequently, the reaction appears to be general for these reactive reagents.

$$(\text{c-C}_5\text{H}_9)_3\text{B} + \text{N}_2\text{CHCHO} \xrightarrow[25^\circ]{\text{THF, H}_2\text{O}} \text{c-C}_5\text{H}_9\text{CH}_2\text{CHO} + \text{N}_2$$

98%

$$(\text{c-C}_6\text{H}_{11})_3\text{B} + \text{N}_2\text{CHCO}_2\text{C}_2\text{H}_5 \xrightarrow[25^\circ]{\text{THF}} \text{c-C}_6\text{H}_{11}\text{CH}_2\text{CO}_2\text{C}_2\text{H}_5$$

58%

The promise of this approach is indicated by its application to achieve the mono- and dialkylation of cyclic ketones.[57]

2-methyl-6-diazocyclohexanone (H$_3$C, O, N$_2$) $\longrightarrow$ 2-methyl-6-ethylcyclohexanone (H$_3$C, O, C$_2$H$_5$) 98%

2,6-bis(diazo)cyclohexanone (N$_2$, O, N$_2$) $\longrightarrow$ 2,6-diethylcyclohexanone (C$_2$H$_5$, O, C$_2$H$_5$) 95%

These reactions are highly promising. They have a major advantage in that they frequently take place very readily at 0° or 25° in the absence of added bases or acids. Therefore, they should be very useful to achieve the functionalization of labile groupings (**P8.15**).

Unfortunately, in its original form the reaction suffers from certain disadvantages. First, the reaction uses but one of the three alkyl groups of the R_3B reactant. Second, the reaction becomes relatively sluggish with sharp decreases in yield with more bulky R groups. For example, the reaction of ethyl diazoacetate with tri(2-methyl-1-butyl)borane is slow, and the yield is only 40%.[55]

Unfortunately, the B-R-9-BBN derivatives do not solve the problem in the present case. They react with ethyl diazoacetate with opening of the bicyclooctyl ring instead of with transfer of the R-group.[58] However, the dialkylchloro-

[57] J. Hooz, D. M. Gunn, and H. Kono, *Can. J. Chem.*, **49**, 2371 (1971).

[58] J. Hooz and D. M. Gunn, *Tetrahedron Lett.*, 3455 (1969); H. C. Brown and M. M. Rogić, *J. Amer. Chem. Soc.*, **91**, 2146 (1969).

boranes **(S3.5)** do provide a considerable improvement. Prepared in ether solution **(P4.14)**, they react directly at −78° to liberate nitrogen and give intermediates, which treated with alcohol at −78°, are converted into the desired products in yields approaching quantitative[59] **(P8.16)**.

$$(\text{c-C}_5\text{H}_9)_2\text{BCl} \xrightarrow[-78^\circ]{N_2CHCO_2Et} N_2 + [\ ] \xrightarrow[-78^\circ]{ROH} \text{c-C}_5\text{H}_9\text{CH}_2\text{CO}_2\text{Et}\quad 94\%$$

The far greater rate of reaction is doubtless a reflection of the smaller steric requirements and higher Lewis acid strength of the R_2BCl molecule as compared to R_3B.

In this reaction one of the two R groups in R_2BCl is still not utilized. This suggested trying the $RBCl_2$ derivatives **(S3.5)**. In the case of the aryldichloroborane, the solution is ideal. Essentially quantitative yields were obtained for several aryl groups.[60]

$$C_6H_5BCl_2 + N_2CHCO_2Et \xrightarrow[-25^\circ,\ -N_2]{THF} [\ ] \xrightarrow{ROH} C_6H_5CH_2CO_2Et$$

The yields are poorer with monoalkyldichloroboranes, in the range of 57-71%, so that for alkyl groups this procedure has little advantage over the more conveniently synthesized reagents, R_2BCl.

The applicability of the R_2BCl and $RBCl_2$ reagents to the other types of diazo derivatives remains to be explored.

7.12. UNSATURATED BORANE TRANSFERS

Thus far this survey has considered only the transfer of alkyl, cycloalkyl, and aryl groups from boron to carbon. It turns out that unsaturated groupings, such as alkenyl (vinyl) and alkynyl, attached to boron can participate in such transfer reactions, both as the migrating group and the migration terminus.

Treatment of the adduct of a dialkylborane and an acetylene with sodium hydroxide and iodine results in the transfer of an alkyl group and the formation of the *cis* olefins[61] **(P8.17)**.

[59] H. C. Brown, M. M. Midland, and A. B. Levy, *J. Amer. Chem. Soc.*, **94**, 3662 (1972).
[60] J. Hooz, J. N. Bridson, J. G. Calzada, H. C. Brown, M. M. Midland, and A. B. Levy, *J. Org. Chem.*, 38, 2574 (1973).
[61] G. Zweifel, H. Arzoumanian, and C. C. Whitney, *J. Amer. Chem. Soc.*, **89**, 3652 (1967).

$$\mathrm{RC{\equiv}CH} \xrightarrow[\mathrm{THF}]{\mathrm{R'_2BH}} \underset{\mathrm{H}}{\overset{\mathrm{R}}{}}\!\!>\!\mathrm{C{=}C}\!<\!\!\underset{\mathrm{BR'_2}}{\overset{\mathrm{H}}{}} \xrightarrow[\mathrm{I_2}]{\mathrm{NaOH}} \underset{\mathrm{H}}{\overset{\mathrm{R}}{}}\!\!>\!\mathrm{C{=}C}\!<\!\!\underset{\mathrm{H}}{\overset{\mathrm{R'}}{}}$$

The reaction is believed to involve (1) coordination of the base with the boron, (2) formation of an iodonium ion, (3) transfer of one of the R′ groups, and (4) *trans*-elimination.

$$\underset{\mathrm{H}}{\overset{\mathrm{R}}{}}\!\!>\!\mathrm{C{=}C}\!<\!\!\underset{\mathrm{BR'_2}}{\overset{\mathrm{H}}{}} \longrightarrow \underset{\mathrm{H}}{\overset{\mathrm{R}}{}}\!\!>\!\mathrm{C{=}C}\!<\!\!\underset{\mathrm{\bar{B}(R')_2{-}OH}}{\overset{\mathrm{H}}{}} \longrightarrow \underset{\mathrm{H}}{\overset{\mathrm{R}}{}}\!\!>\!\mathrm{C{-}\overset{\overset{+}{I}}{}{-}C}\!<\!\!\underset{\mathrm{B(R')_2{-}OH}}{\overset{\mathrm{H}}{}} \longrightarrow$$

$$\mathrm{I{-}C(R)(H){-}C(H)(R')(B(R'){-}OH)} \longrightarrow \mathrm{I{-}C(R)(H){-}C(R')(H)\cdots BROH} \longrightarrow \underset{\mathrm{H}}{\overset{\mathrm{R}}{}}\!\!>\!\mathrm{C{=}C}\!<\!\!\underset{\mathrm{H}}{\overset{\mathrm{R'}}{}}$$

The Zweifel *cis*-olefin synthesis is a major development. It requires a simple route to dialkylboranes, now under development, and a means of avoiding the loss of 50% of the R′ groups in R'_2BH for cases where R′ represents a valuable intermediate. Unfortunately, the thexylmonoalkylboranes **(S3.6)** do not solve the problem–both groups migrate in the corresponding adduct.[62]

The Zweifel *trans*-olefin synthesis involves an ingenious modification.[63]

$$\mathrm{RC{\equiv}CBr} \xrightarrow{\mathrm{R'_2BH}} \underset{\mathrm{H}}{\overset{\mathrm{R}}{}}\!\!>\!\mathrm{C{=}C}\!<\!\!\underset{\mathrm{BR'_2}}{\overset{\mathrm{Br}}{}} \xrightarrow{\mathrm{NaOCH_3}} \underset{\mathrm{H}}{\overset{\mathrm{R}}{}}\!\!>\!\mathrm{C{=}C}\!<\!\!\underset{\mathrm{R'}}{\overset{\mathrm{BR'OCH_3}}{}} \xrightarrow{\mathrm{RCO_2H}} \underset{\mathrm{H}}{\overset{\mathrm{R}}{}}\!\!>\!\mathrm{C{=}C}\!<\!\!\underset{\mathrm{R'}}{\overset{\mathrm{H}}{}}$$

It is of interest that base produces a *transfer* with inversion even though the halogen atom is vinylic. Fortunately, in this case the thexylmonoalkylboranes can be utilized.[64] **(P8.18)**.

[62]J.-J. Katz, Ph. D. thesis, Purdue University.

[63]G. Zweifel and H. Arzoumanian, *J. Amer. Chem. Soc.*, **89**, 5086 (1967); G. Zweifel, R. P. Fisher, J. T. Snow, and C. C. Whitney, *J. Amer. Chem. Soc.*, **93**, 6309 (1971).

[64]E. Negishi, J.-J. Katz, and H. C. Brown, *Synthesis*, 555 (1972).

$$R{-}C{\equiv}CBr \xrightarrow{\text{(thexyl)}BHR'} \begin{matrix} R & & Br \\ & C{=}C & \\ H & & B(\text{thexyl})R' \end{matrix} \xrightarrow{NaOCH_3} \begin{matrix} R & & B(\text{thexyl})OCH_3 \\ & C{=}C & \\ H & & R' \end{matrix}$$

$$\xrightarrow{RCO_2H} \begin{matrix} R & & H \\ & C{=}C & \\ H & & R' \end{matrix}$$

A wide variety of structures of R′ gave yields of *trans*-olefins in the 90% range.
The Zweifel synthesis has been extended to the preparation of *cis,trans*-dienes.[65]

$$2\,RC{\equiv}CH + (\text{thexyl})BH_2 \longrightarrow [(R)(H)C{=}C(H)]_2B(\text{thexyl}) \xrightarrow[2.\ H_2O]{1.\ (CH_3)_3NO} [(R)(H)C{=}C(H)]_2B{-}OH \xrightarrow[I_2]{NaOH} \begin{matrix} R & & H & & \\ & C{=}C & & & R \\ H & & C{=}C & & \\ & H & & & H \end{matrix}$$

This procedure can be considerably shortened by utilizing chloroborane for the hydroboration stage[66] (**S3.5, P8.19**).

$$2\,RC{\equiv}CH \xrightarrow{ClBH_2} [(R)(H)C{=}C(H)]_2BCl \xrightarrow[I_2]{NaOH} \begin{matrix} R & & H & & \\ & C{=}C & & & R \\ H & & C{=}C & & \\ & H & & & H \end{matrix}$$

[65]G. Zweifel, N. L. Polston, and C. C. Whitney, *J. Amer. Chem. Soc.*, **90**, 6243 (1968).
[66]H. C. Brown and N. R. Ravindran, *J. Org. Chem.*, **38**, 1617 (1973).

More recently Negishi has achieved a general synthesis of *trans,trans*-dienes[67] **(P8.20)**.

$$RC{\equiv}CCl \xrightarrow{\text{(disiamyl)}BH_2} \text{(Z)-RCH=CCl–B(disiamyl)} \xrightarrow{R'C{\equiv}CH} \text{RCH=CCl–B(–CH=CHR')(disiamyl)}$$

$$\xrightarrow{NaOCH_3} \text{RCH=C(B(OCH}_3\text{)(disiamyl))–CH=CHR'} \xrightarrow{RCO_2H} \text{RCH=CH–CH=CHR'}$$

Finally, the "ate" complexes of acetylenes can be utilized in various ways in a rapidly developing area. Thus they provide a remarkably simple versatile route to both mono-[68] **(P8.22)** and disubstituted acetylenes[69] **(P8.21)**.

$$(C_6H_5)_3B \xrightarrow{LiC{\equiv}C{-}C(CH_3)_3} \left[(C_6H_5)_3B{-}C{\equiv}C{-}C(CH_3)_3\right]Li \xrightarrow[-78^\circ]{I_2} C_6H_5{-}C{\equiv}C{-}C(CH_3)_3 \quad 94\%$$

Protonolysis or alkylation of these "ate" complexes, followed by oxidation, provides a simple route to a wide variety of ketones.[70]

[67]E. Negishi and T. Yoshida, *J. Chem. Soc. Chem. Commun.*, 606 (1973).
[68]M. M. Midland, J. A. Sinclair, and H. C. Brown, *J. Org. Chem.*, **39**, 731 (1974).
[69]A. Suzuki, N. Miyaura, S. Abiko, M. Itoh, H. C. Brown, J. A. Sinclair, and M. M. Midland, *J. Amer. Chem. Soc.*, **95**, 3080 (1973).
[70]A. Pelter, C. R. Harrison, and D. Kirkpatrick, *J. Chem. Soc. Chem. Commun.*, 544 (1973).

$$R_3BC{\equiv}CR' \xrightarrow{HX} \underset{R_2B}{\overset{R}{\diagdown}}C{=}C\underset{H}{\overset{R'}{\diagup}} \xrightarrow{[O]} RCOCH_2R'$$

(*cis, trans* isomers)

7.13. CONJUGATE ADDITIONS

Simple trialkylboranes do not undergo 1,2 addition to the carbonyl groups of aldehydes and ketones. However, they do undergo a facile 1,4 addition to α,β-unsaturated aldehydes[71] and ketones.[72]

$$R_3B + CH_2{=}CHCHO \xrightarrow[25^\circ]{THF} RCH_2CH{=}CHOBR_2 \xrightarrow{H_2O} RCH_2CH_2CHO$$

$$R_3B + CH_2{=}CHCOCH_3 \xrightarrow[25^\circ]{THF,\ H_2O} RCH_2CH_2COCH_3$$

Clearly this provides a simple means of lengthening the chain by three or four carbon atoms.

Many substituents can be accommodated. Of especial interest is the simple synthesis of α-bromoaldehydes afforded by the use of 2-bromoacrolein[73] (**P8.23**).

$$sec\text{-}Bu_3B + CH_2{=}C(Br){-}CHO \xrightarrow[25^\circ]{THF,\ H_2O} CH_3CH_2CH(CH_3)CH_2CH(Br)CHO$$

81%

Mannich bases derived from ketones such as cyclopentanone, cyclohexanone, and norcamphor, quaternized *in situ*, react smoothly in alkaline solution with the organoboranes.[74]

[71]H. C. Brown, M. M. Rogić, M. W. Rathke, and G. W. Kabalka, *J. Amer. Chem. Soc.*, **89**, 5709 (1967).

[72]A. Suzuki, A. Arase, H. Matsumoto, M. Itoh, H. C. Brown, M. M. Rogić, and M. W. Rathke, *J. Amer. Chem. Soc.*, 89, 5708 (1967).

[73]H. C. Brown, G. W. Kabalka, M. W. Rathke, and M. M. Rogić, *J. Amer. Chem. Soc.*, **90**, 4165 (1968).

[74]H. C. Brown, M. W. Rathke, G. W. Kabalka, and M. M. Rogić, *J. Amer. Chem. Soc.*, **90**, 4166 (1968).

$$\text{cyclopentanone} \longrightarrow \text{2-}(CH_2N(CH_3)_2\cdot HCl)\text{cyclopentanone} \xrightarrow[K_2CO_3]{CH_3I,\ Et_3B} \text{2-}(CH_2CH_2CH_3)\text{cyclopentanone}\ \ 90\%$$

In some cases the reaction appears sluggish. In such cases, a small amount of air induces a rapid reaction.

$$\text{cyclohexenone} + (\text{cyclohexyl})_3B \xrightarrow{THF,\ H_2O} \text{No reaction}$$

$$\xrightarrow{\text{cat. } O_2} \text{3-cyclohexylcyclohexanone}$$

The reaction involves an interesting type of chain reaction.

$$R_3B \xrightarrow{O_2} R\cdot$$

$$R\cdot + CH_2{=}CHCHO \longrightarrow RCH_2\dot{C}HCHO \leftrightarrow RCH_2CH{=}CHO\cdot$$

$$RCH_2CH{=}CHO\cdot + R_3B \longrightarrow RCH_2CH{=}CHOBR_2 + R\cdot$$

In the absence of the organoborane, free radicals would initiate the usual type of vinyl polymerization. Organoboranes, however, are enormously reactive toward free radicals containing an unpaired electron on oxygen or nitrogen.[75] Consequently, these capture the free radical intermediates and continue the chain.

In these reactions, only one of the three R groups of R_3B are used. Fortunately, this problem can be circumvented here by the use of the 3,5-dimethylborinane derivatives **(P4.8)** **(P8.24)** or alkyldiphenylboranes.[76]

[75] K. U. Ingold and B. P. Roberts, *Free Radical Substitution Reactions*, Interscience, New York, 1971.

[76] Research in progress with P. Jacob.

$$\text{(B-thexyl borinane)} \xrightarrow[\text{H}_2\text{O, cat. O}_2]{\text{2-cyclohexenone}} \text{3-}C(CH_3)_3\text{-cyclohexanone} \quad 73\%$$

$$\text{(B-alkyl borinane)} \xrightarrow[\text{H}_2\text{O, cat. O}_2]{CH_2{=}CHCOCH_3} (H_3C)_2HC{-}C(CH_3)_2CH_2CH_2COCH_3 \quad 88\%$$

The products of these reactions are enol borinates.[71,72]

$$(C_2H_5)_3B + CH_2{=}CHCOCH_3 \longrightarrow CH_3CH_2CH_2CH{=}C(OB(C_2H_5)_2)CH_3$$

Similar enol borinates are formed in the reaction of diazo derivatives with organoboranes.[77]

$$(C_2H_5)_3B + N_2CHCOCH_3 \longrightarrow CH_3CH_2CH{=}C(OB(C_2H_5)_2)CH_3 + N_2$$

Such enol borinates have interesting possibilities in synthesis.[78] For example, treatment of such enol borinates with *N*-bromosuccinimide[79] or dimethyl (methylene) ammonium iodide[80] results in a highly useful regiospecific synthesis of α-haloketones and Mannich bases.

$$CH_3CH_2CH_2CH{=}C(OB(C_2H_5)_2)CH_3 \xrightarrow{\text{NBS}} CH_3CH_2CH_2CH(Br)COCH_3$$

$$CH_3CH_2CH_2CH{=}C(OB(C_2H_5)_2)CH_3 \xrightarrow{(CH_3)_2\overset{+}{N}{=}CH_2I^-} CH_3CH_2CH_2CH(CH_2N(CH_3)_2)COCH_3$$

[77] D. J. Pasto and P. W. Wojtkowski, *Tetrahedron Lett.*, 215 (1970).
[78] T. Mukaiyama, K. Inomata, and M. Muraki, *J. Amer. Chem. Soc.*, **95**, 967 (1973).
[79] J. Hooz and J. N. Bridson, *Can. J. Chem.*, **50**, 2387 (1972).
[80] J. Hooz and J. N. Bridson, *J. Amer. Chem. Soc.*, **95**, 602 (1973).

In this brief survey, it has been possible to cover only a few of the ramifications of the 1,4-addition reaction and of the organoboranes as a source of free radicals. More extensive discussions are available.[1,2,75,81]

7.14. CONCLUSION

It is only a few years since the utility of organoboranes for the formation of carbon-carbon bonds was first recognized. However, the last few years have witnessed a burst of developments in this area that is probably unprecedented in any previously uncovered synthetic process. Clearly, the organic chemist interested in applying the best, most elegant procedure to resolve a problem in synthesis must become familiar with and proficient in the use of these new reactions.

[81] H. C. Brown and M. M. Midland, *Angew. Chem. Internat. Edit. Engl.*, **11**, 692 (1972).

8

CARBON BOND FORMATION VIA ORGANOBORANES: PROCEDURES

8.1. DOCOSANEDIOIC ACID

$$3\ CH_3O_2C(CH_2)_8CH{=}CH_2 + BH_3{:}THF \xrightarrow[THF]{0°} [CH_3O_2C(CH_2)_8CH_2CH_2]_3B$$

$$2\,[CH_3O_2C(CH_2)_8CH_2CH_2]_3B + 6\ AgNO_3 + 14\ KOH \xrightarrow[CH_3OH]{25°}$$

$$3\ KO_2C(CH_2)_{20}CO_2K + 6\ Ag + 6\ KNO_3 + 2\ KB(OH)_4 + 6\ CH_3OH$$

$$KO_2C(CH_2)_{20}CO_2K + 2\ HCl \longrightarrow HO_2C(CH_2)_{20}CO_2H + 2\ KCl$$

Procedure by M. K. Unni[1]

Procedure

A dry 500-ml flask equipped with a septum inlet, pressure-equalizing dropping funnel, magnetic stirring bar, and a reflux condenser connected to a mercury bubbler is flushed with nitrogen and then maintained under static nitrogen pressure **(S9.1)** until after the coupling stage. The flask is charged with 19.8 g (100 mmole) of methyl 10-undecenoate **(S9.11)**. The flask is immersed in an ice-water bath and hydroboration achieved by the dropwise addition of 33.4 ml (33.4 mmoles) of a 1.0*M* solution of borane-tetrahydrofuranate to the ester with vigorous stirring **(N1)**. After 2 hours at 0°, 20 ml of methanol is added to destroy any residual hydride. Then 120 ml of 2.0*M* potassium hydroxide in methanol is introduced with the aid of a double-ended needle (**S9.2**). This is followed by addition over 10 minutes of 24 ml of 5.0*M* aqueous silver nitrate to the vigorously stirred solution maintaining the temperature at 20 to 30°. The protective atmosphere is now no longer necessary. After 1 hour, 30 g of potassium hydroxide is added and the resulting mixture refluxed for 3 hours to hydrolyze the coupled products. The mixture is poured into 300 ml of water, acidified with concentrated hydrochloric acid to pH2, filtered, and the filtrate discarded. The gray, sticky precipitate containing the silver residues is extracted with acetone (Soxhlet), and the extract is cooled to 0° (**N2**). The dicarboxylic acid is removed by filtration. Recrystallization from acetone yields 8.15 g (44% yield) of docosanedioic acid as white needles, mp 125° [lit.[2] 125-125.5°].

Notes

1. With olefins containing reactive functional groups such as ester, keto, nitrile, etc., stoichiometric amount of borane is utilized (1 mole of borane for 3 moles of the olefin) to avoid any concurrent reduction of the functional group.
2. The silver is removed by dissolving the free acid in boiling acetone rather than filtering a hot solution of the potassium salt of the dicarboxylic acid because even in boiling water this salt is relatively insoluble.[3]

[1] H. C. Brown and C. H. Snyder, *J. Amer. Chem. Soc.*, **83**, 1001 (1961). M. K. Unni, unpublished research.
[2] J. P. Riley, *J. Chem. Soc.*, 2193 (1953).
[3] N. L. Drake, H. W. Carhart, and R. Mozingo, *J. Amer. Chem. Soc.*, **63**, 617 (1941).

8.2. CYCLOPROPYLCARBINYL CHLORIDE

$$\text{(C)BH} + CH_2{=}CHCHClCH_2Cl \xrightarrow[\text{THF}]{25^\circ} \text{(C)}BCH_2CH_2CHClCH_2Cl$$

$$\text{(C)}BCH_2CH_2CHClCH_2Cl + NaOH \xrightarrow[\text{THF-}H_2O]{25^\circ} \triangleright\text{-}CH_2Cl + \text{(C)}BOH + NaCl$$

Procedure by Stanley P. Rhodes[1]

Procedure

A dry 500-ml flask equipped with a septum inlet, a pressure-equalizing dropping funnel, a magnetic stirring bar, and a mercury bubbler is assembled, flushed with nitrogen, and maintained under a static pressure of the gas until following the cyclization of the intermediate with base **(S9.1)**. In the flask is placed 200 ml of 0.5*M* 9-BBN (0.100 mole) in tetrahydrofuran **(P2.11** or Aldrich). Then 12.5 g (0.100 mole) of 3,4-dichloro-1-butene **(N1)** is added neat over 10 minutes, and the reaction mixture is stirred at 25° for 1 hour. Then 40 ml of 3*M* aqueous sodium hydroxide (0.120 mole) is added dropwise. After 2 hours the flask is opened, anhydrous potassium carbonate is added to saturate the aqueous phase, and the upper tetrahydrofuran layer is separated. After drying over anhydrous magnesium sulfate, VPC analysis indicates a yield of 81%. Distillation gives 6.3 g (70% yield) of cyclopropylcarbinyl chloride, bp 83-90° (742 mm), $n^{20}\text{D}$ 1.4360 identical in NMR spectrum with that reported[2] **(N2)**.

Notes

1. Petro-Tex Chemical Corp., Houston, Texas, or Aldrich Chemical Company **(S9.11)**.
2. The same procedure converts 2-phenylallyl chloride into phenylcyclopropane in a yield of 92% and 1,1-dichloro-2-propene into cyclopropyl chloride in a yield of 90%.

[1] H. C. Brown and S. P. Rhodes, *J. Amer. Chem. Soc.*, **91**, 2149 (1969).
[2] M. C. Caserio, W. H. Graham, and J. D. Roberts, *Tetrahedron*, **11**, 171 (1960).

8.3. TRI-2-NORBORNYLCARBINOL

$$12 \quad + 3\,NaBH_4 + 4\,BF_3 \xrightarrow[0^\circ]{DG} 4 \quad)_3B \quad + 3\,NaBF_4$$

$$)_3B \quad + CO + (CH_2OH)_2 \xrightarrow[100\text{-}150^\circ]{DG} 4 \quad)_3CB\left<\begin{matrix}O-CH_2\\ O-CH_2\end{matrix}\right. + H_2O$$

$$(\text{C}_7\text{H}_{11})_3\text{CB}\begin{matrix}\text{O-CH}_2\\ |\\ \text{O-CH}_2\end{matrix} + H_2O_2 + NaOH \xrightarrow[50^\circ]{DG} (\text{C}_7\text{H}_{11})_3\text{COH} + NaB(OH)_4 + (CH_2OH)_2$$

Procedure by Michael W. Rathke [1]

Procedure

A dry 500-ml flask equipped with a septum inlet, thermometer well, and magnetic stirring bar is attached to the automatic gasimeter set up for carbon monoxide delivery **(S9.4)**. The system is flushed with nitrogen and maintained under a static pressure of the gas **(S9.1)**. The flask is charged with 75 ml of a 1.00*M* solution of sodium borohydride (0.075 mole) in diglyme and 28.3 g (0.300 mole) of norbornene in 75 ml of diglyme. The flask is immersed in an ice-water bath and hydroboration carried out by the dropwise addition of 27.4 ml (0.100 mole) of 3.65*M* boron trifluoride diglymate **(S9.11)** to the stirred solution. After 1 hour at room temperature, 10 ml of ethylene glycol is added, and the flask and its contents are brought to 100°. The stirring is stopped, and the system is flushed with carbon monoxide **(S9.4)**. The absorption of carbon monoxide by the organoborane solution is initiated by vigorous stirring of the reaction mixture. After 1 hour, the absorption is complete **(N1)**. The system is flushed with nitrogen and the reaction solution heated to 150° for 1 hour to ensure completion of the migration stage. The reaction mixture is then brought to room temperature, 33 ml of 6*M* sodium hydroxide (100% excess) added, the flask immersed in a water bath, and the oxidation carried out by the dropwise addition of 33 ml of 30% hydrogen peroxide (100% excess) **(N2)**, maintaining a temperature of just under 50°. The solution is then maintained at 50° for 3 hours to complete the oxidation. Addition of the cooled solution to 300 ml of water results in the precipitation of the product. The solid is collected on a filter and crystallized from pentane. There is obtained 25 g (80% yield) of tri-2-norbornylcarbinol, mp 137-137.5° **(N3)**.

Notes

1. The absorption is 50% of theoretical in 10 minutes. Other organoboranes may require longer or shorter reaction times.[1] The reaction should be carried on until CO absorption has stopped.
2. This is an example of a difficult oxidation **(P6.10)**. The oxidation of 0.100 mole of a compound R_3CBO would normally require only 33 ml of 3*M* sodium hydroxide and 11 ml of 30% hydrogen peroxide.

3. By analogy with many other related migrations of groups from boron to carbon (**S7**), the product is doubtless tri-*exo*-norbornylcarbinol. However, the stereochemistry of the bond at the 2-position has not been unambiguously established. Other possibilities for stereoisomerism exist. Consequently, the stereochemically noncommittal name, tri-2-norbornylcarbinol, is used.

[1] H. C. Brown and M. W. Rathke, *J. Amer. Chem. Soc.*, 89, 2737 (1967).

8.4. ETHYL 7-METHYL-5-OXOOCTANOATE

$$\text{thexyl-}BH_2 + H_2C{=}C(CH_3)_2 \xrightarrow[-20^\circ]{THF} \text{thexyl-}B(H)CH_2CH(CH_3)_2$$

$$\text{thexyl-}B(H)CH_2CH(CH_3)_2 + H_2C{=}CHCH_2CO_2C_2H_5 \xrightarrow[0^\circ]{THF} \text{thexyl-}B(CH_2CH_2CH_2CO_2C_2H_5)(CH_2CH(CH_3)_2)$$

$$\text{thexyl-}B(CH_2CH_2CH_2CO_2C_2H_5)(CH_2CH(CH_3)_2) + CO + 2\,H_2O \xrightarrow[70\text{ atm}]{THF,\ 50^\circ} \text{thexyl-}B(OH){-}C(OH)(CH_2CH_2CH_2CO_2C_2H_5)(CH_2CH(CH_3)_2)$$

$$\text{thexyl-}B(OH){-}C(OH)(CH_2CH_2CH_2CO_2C_2H_5)(CH_2CH(CH_3)_2) \xrightarrow[NaOAc]{H_2O_2} (CH_3)_2CHCH_2COCH_2CH_2CH_2CO_2C_2H_5$$

Procedure by Ei-ichi Negishi[1]

Procedure

A dry 300-ml round-bottomed flask, equipped with a septum inlet, magnetic stirring bar, and mercury bubbler, is flushed with nitrogen and maintained under a static pressure of nitrogen (**S9.1**). The flask is immersed in an ice-salt bath, and a solution of thexylborane in THF, 100 ml of 1.00*M* (0.100 mole), is prepared following the procedure of **P2.10**. Isobutylene from a cylinder is condensed into a calibrated tube (cooled in a $-10°$ bath), and 8.9 ml (0.100 mole) is measured out and transferred by volatilization into the thexylborane solution (**S9.3**). After 30 minutes at 0°, 11.4 g (0.100 mole) of ethyl vinylacetate (**S9.11**) is added. The mixture is allowed to stir for 2 hours at 0°. Then 3.6 ml of water (0.200 mole) is added. A 200-ml stainless steel autoclave is capped by a two-hole rubber stopper containing two glass tubes each fitted with rubber serum caps. The autoclave is flushed with nitrogen by means of a stream flowing in and out of the two serum caps penetrated by hypodermic needles. The solution is now transferred to the autoclave under nitrogen with the aid of a double-ended needle. Small amounts of tetrahydrofuran (2 × 10 ml) are used to wash the flask and transferred to the autoclave. The autoclave cap is fastened and carbonylation carried out by filling the autoclave with carbon monoxide to 70 atm (**N1**), maintaining the temperature at 50° for 3 hours (**N2**). The autoclave is brought to room temperature and the solution transferred to a glass flask (**N3**). Oxidation is carried out by adding 33 ml of 3*M* sodium acetate followed by 24 ml of 30% hydrogen peroxide (exothermic!, **P2.1:N6**), controlling the rate of addition so as to maintain the temperature in the range of 30-40°. The reaction mixture is then maintained for 1 hour at 50° to ensure completion of the oxidation. After cooling, the aqueous layer is saturated with sodium chloride and the organic phase separated. The aqueous phase is extracted with ether (2 × 50 ml). The combined extracts are dried over anhydrous magnesium sulfate and distilled. There is obtained 16.1 g (81% yield) of ethyl 7-methyl-5-oxooctanoate, bp 102-104° (4 mm), n^{20}D 1.4326, essentially pure by VPC and NMR examination.

Notes

1. In contrast to tricyclohexylborane and tri-*exo*-norbornylborane (**P8.3**), these thexyl derivatives take up carbon monoxide only very slowly at atmospheric pressure.
2. Actually, the absorption of carbon monoxide is usually complete in 1 hour at 50°. The reactants are maintained at 50° for 3 hours to ensure completion.
3. No special precautions to protect the solution from the atmosphere are essential at this point.

[1] H. C. Brown and E. Negishi, *J. Amer. Chem. Soc.*, 89, 5285 (1967).

8.5. 11-CARBOMETHOXYUNDECANAL

$$\text{(9-BBN)BH} + CH_2{=}CH(CH_2)_8CO_2CH_3 \xrightarrow[25^\circ]{THF} \text{(9-BBN)}BCH_2CH_2(CH_2)_8CO_2CH_3$$

$$\text{(9-BBN)}B(CH_2)_{10}CH_2CH_3 + Li(t\text{-}BuO)_3AlH + CO \xrightarrow[-30^\circ]{THF} \text{(9-BBN)}B{-}\overset{H}{\underset{Li(t\text{-}BuO)_3AlO}{C}}{-}(CH_2)_{10}CO_2CH_3$$

$$\text{(9-BBN)}B{-}\overset{H}{\underset{Li(t\text{-}BuO)_3AlO}{C}}{-}(CH_2)_{10}CO_2CH_3 + 3\,H_2O_2 \xrightarrow[25^\circ]{pH\,7} H\overset{O}{\overset{\|}{C}}(CH_2)_{10}CO_2CH_3$$

Procedure by Randolph A. Coleman[1]

Procedure

The automatic gasimeter is set up for carbon monoxide generation (**S9.4**). The dried 500-ml reaction vessel of the gasimeter, fitted with a septum inlet and a magnetic stirring bar, is flushed with nitrogen (**S9.1**). The flask is charged with 20 ml of THF and 9.9 g (0.050 mole) of methyl 10-undecenoate (**P6.7:N1**). Hydroboration is achieved by adding slowly (over 20 minutes) 100 ml of a 0.5*M* solution of 9-BBN in tetrahydrofuran (**P2.11** or Aldrich). The reaction mixture is permitted to stir for 1 hour at 25° to complete the hydroboration. The flask is cooled to −25 to −35° (**N1**), the system is flushed with carbon monoxide, and reaction carried out by adding 50 ml of a 1.0*M* solution of lithium tri-*t*-butoxyaluminohydride[2] over 45 minutes (**N2**). Complete absorption of carbon monoxide requires an additional 90 minutes. The reaction mixture is brought to room temperature, 120 ml of a pH 7 buffer is added (a water solution 2.2*M* in NaH_2PO_4 and K_2HPO_4), and oxidation is carried out by the dropwise addition of 21 ml of 30% hydrogen peroxide, keeping the temperature below 30°. The tetrahydrofuran phase is washed twice with a saturated sodium chloride-water solution and then stirred for 1 hour with 100 ml of a saturated aqueous sodium bisulfite solution. The precipitated adduct is washed with pentane and then treated with 50 ml of a saturated aqueous magnesium sulfate solution, 50 ml of pentane, and 5 ml of a 40% solution of formaldehyde (**N3**). After the bisulfite adduct has disappeared (~30 minutes), the pentane layer is removed. Distillation yields 9.7 g (85% yield) of 11-carbomethoxyundecanal, bp 115-118° (1 mm), n^{20}D 1.4452, mp 2,4-DNP 70-71.5°.

Notes

1. Previous procedures utilized lithium trimethoxyaluminum hydride and temperatures of 0°.[3] Lithium tri-*t*-butoxyaluminohydride[2] is a milder reducing agent than lithium trimethoxyaluminohydride and consequently is advantageous in applying this reaction to olefins containing functional groups capable of undergoing reduction.[1] The lower temperature enhances the migration of the B–R group and diminishes that of the cyclooctyl moiety.
2. It is important to follow the procedure closely, since lithium tri-*tert*-butoxyaluminohydride reacts with organoboranes to produce intermediates which can reductively open the ring of tetrahydrofuran and other cyclic and bicyclic ethers.[4]
3. This procedure for the regeneration of aldehyde from the bisulfite adduct proved to be the best of those examined.[5] The time required to achieve the essentially complete precipitation of the bisulfite adduct can vary considerably in individual preparations.

[1] H. C. Brown and R. A. Coleman, *J. Amer. Chem. Soc.*, **91**, 4606 (1969).
[2] H. C. Brown and P. M. Weissman, *Israel J. Chem.*, **1**, 430 (1963).
[3] H. C. Brown, E. F. Knights, and R. A. Coleman, *J. Amer. Chem. Soc.*, **91**, 2144 (1969).
[4] H. C. Brown, S. Krishnamurthy, and R. A. Coleman, *J. Amer. Chem. Soc.*, **94**, 1750 (1972).
[5] J. W. Cornforth, *Org. Syn.*, **Coll. Vol. 4**, 757 (1963).

8.6. *TRANS*-1-HYDRINDANONE

BH_2

H B H

H B H

CO/70 atm / H_2O → H_2O_2 / NaOAc →

H H O

Procedure by Ei-ichi Negishi[1]

Procedure

A dry 300-ml three-neck flask, equipped with a septum inlet, magnetic stirring bar, mercury bubbler, and two pressure-equalizing dropping funnels is flushed with nitrogen and maintained under a static pressure of the gas (**S9.1**). In the flask is placed 25 ml of tetrahydrofuran (**S9.11**). Thexylborane, 50 ml of a 1.00*M* solution (**P2.10**), is introduced into one dropping funnel with a syringe and 5.95 g (0.055 mole) of 1-vinylcyclohexene (**N1**) in 25 ml of tetrahydrofuran into the other. The solution in the flask is stirred vigorously as the two solutions are added simultaneously at approximately equivalent rates over 3 hours at 25°. The solution is stirred for an additional 5 hours to complete the cyclization (**N2**). Then 1.8 ml (0.100 mole) of water is added and the mixture transferred into a 250-ml autoclave under nitrogen (**P8.4**) and carbonylated at 1000 psi and 50° (**P8.4:N1**) for 3 hours (**P8.4:N2**). The reaction mixture is transferred to a glass flask and oxidized with 20 ml of 3*M* sodium acetate and 20 ml of 30% hydrogen peroxide (exothermic! **P2.1:N6**) at 30-50°. The reaction mixture is maintained at 50° for 1 hour to complete the oxidation. The material is cooled, the aqueous phase saturated with sodium chloride, and the tetrahydrofuran phase separated. VPC analysis indicates a yield of 66%. Distillation provides 3.7 g (54% yield) of *trans*-1-hydrindanone, bp 40-43° (0.5 mm), n^{20}D 1.4782, mp oxime 145-146° [lit.,[3] mp 146°] (**N3**).

Notes

1. 1-Vinylcyclohexanol, from the vinyl Grignard in tetrahydrofuran and cyclohexanone, is dehydrated with potassium acid phthalate, bp 52-53° (46 mm), n^{20}D 1.4637 (**S9.11**).
2. For most cyclic hydroborations, as with 1,4-pentadiene and 1,5-hexadiene, the reactions are carried out at 0° for much shorter reaction periods. In the present case the cyclization appears to be considerably slower, presumably a consequence of the formation of a relatively rigid bicyclic.
3. This annelation reaction appears to be broadly applicable (**S7.4**).

[1] H. C. Brown and E. Negishi, *J. Amer. Chem. Soc.*, 89, 5477 (1967).
[2] H. C. Brown and E. Negishi, *Chem. Commun.*, 594 (1968).
[3] W. Hückel, M. Sachs, J. Yantschulewitsch, and F. Nerdel, *Ann.*, **518**, 155 (1935).

8.7. PERHYDRO-9b-PHENALENOL (TERCYCLANOL)

$$+ H_3B{:}N(C_2H_5)_3 \xrightarrow[140^\circ]{DG} \text{B} + (C_2H_5)_3N$$

$$\text{perhydro-9b-boraphenalene} + CO + (CH_2OH)_2 \xrightarrow{150^\circ} \text{B-(ethylenedioxy)boryl derivative}$$

$$\text{B-(ethylenedioxy)boryl derivative} + H_2O_2 + 2\,H_2O + NaOH \xrightarrow[50^\circ]{EtOH} \text{perhydro-9b-phenalenol (OH)}$$

$$+ NaB(OH)_4 + (CH_2OH)_2$$

Procedure by Ei-ichi Negishi and William C. Dickason[1]

Procedure

A 1-ℓ flask fitted with a septum inlet, a magnetic stirring bar, a short Vigreux column connected to a distillation setup **(S9.7)**, and a mercury bubbler is assembled, flushed with nitrogen, and then maintained under a static pressure of the gas **(S9.1)**. In the flask is placed 500 ml (0.500 mole) of 1.00*M* borane-tetrahydrofuran complex **(P2.3, P2.4:N3)**. The flask is immersed in an ice-water bath, and the borane-triethylamine complex is prepared *in situ* by the slow addition of 50.6 g (0.500 mole) of triethylamine **(N1)**. Most of the tetrahydrofuran is removed by distillation at atmospheric pressure. Pure, dry diglyme **(S9.11)**, 300 ml, is now added. The temperature of the reaction mixture is now maintained at 130-140° as a solution of 81 g (0.500 mole) of *trans,trans,trans*-1,5,9-cyclododecatriene **(N2)** **(S9.11)** in 100 ml of digylme is added over 2 hours by means of a syringe pump **(N3)**. Most of the diglyme is removed by distillation at atmospheric pressure, and the product is heated at 200° for 6 hours **(N4)**. After cooling, the thermally treated product is transferred under nitrogen to a smaller distillation setup, and a small amount of tetrahydrofuran is used to achieve a quantitative transfer. Distillation provides 74.6 g (85% yield) of a 92:8 mixture of *cis,trans*- and all-*cis*-perhydro-9*b*-boraphenalene, bp 115-117° (10 mm). Distillation in an efficient spinning band column provides *cis,trans*-perhydro-9*b*-boraphenalene in a purity of 98% (VPC analysis).

A 250-ml autoclave **(N5)** is closed with a two-hole rubber stopper carrying two glass tubes fitted with rubber septums. The autoclave is flushed with nitrogen introduced through a hypodermic needle penetrating one septum and led away to a mercury bubbler through a needle penetrating the second septum. The autoclave is then maintained under a static pressure of nitrogen as the autoclave is charged with 17.6 g (0.100 mole) of *cis,trans*-perhydro-9*b*-

boraphenalene, 50 ml of tetrahydrofuran (**S9.11**), and 18.6 g (16.8 ml, 0.300 mole) of ethylene glycol. The rubber stopper is now removed in a stream of nitrogen, and the top of the autoclave is attached to the autoclave with minimum exposure of the contents. The autoclave is now pressured with carbon monoxide from a cylinder to a pressure of 1000 psi. The autoclave temperature is raised to 150° and maintained there for 2 hours. The apparatus is brought to room temperature and opened to the air. The reaction mixture is diluted with 100 ml of pentane, washed with water, and dried over magnesium sulfate. The solution is filtered. Removal of the solvents under reduced pressure provides crude *cis,cis,trans*-2-(perhydro-9′*b*-phenalyl)-1,3,2-dioxaborole in nearly quantitative yield. Recrystallization from pentane at −78° provides a pure sample which melts sharply at 102-102.5° (**N6**).

A solution of 24.8 g (0.100 mole) of *cis,cis,trans*-2-(perhydro-9′*b*-phenalyl)-1,3,2-dioxaborole in 100 ml of tetrahydrofuran and 100 ml of 95% ethanol is stirred with 37 ml (120% excess) of 6*M* sodium hydroxide (0.200 mole, 120% excess) and then 37 ml of 30% hydrogen peroxide (120% excess) is added dropwise at a rate such that the temperature does not exceed 40° (**N7**). After the original reaction has subsided, the reaction mixture is maintained at 50° for 3 hours to complete the reaction. The reaction mixture is brought to 25°, an equal volume of pentane is added, and the organic phase separated. The organic phase is washed with water (3 × 50 ml) and then dried over magnesium sulfate. Removal of the pentane produces 13.6 g of pure *cis,cis,trans*-perhydro-9*b*-phenalenol, mp 78-78.5° (from pentane), a yield of 70% (**N8, N9**).

Notes

1. The direct utilization of borane-tetrahydrofuran for the hydroboration results in the formation of a polymeric intermediate. This product can be depolymerized by heating. This difficulty is avoided by the use of borane-triethylamine.[2]
2. Other isomers, such as the *trans,trans,cis*-1,5,9-cyclododecatriene, or a mixture of isomers, can be used. In that case, slightly different isomer distributions are observed. The *trans,trans,trans* isomer appears to provide the maximum formation of the desired *cis,cis,trans*-perhydro-9*b*-phenalenol.
3. Alternatively, a pressure-equalizing dropping funnel mounted on top of the Vigreux column can be used (**P2.11**).
4. It is essential to maintain the temperature of the product (not the bath temperature) at 200° to achieve the isomerization of the other constitutional isomers to perhydro-9*b*-boraphenalenes. The present procedure is largely based on that described by Rotermund and Köster.[2] However, these authors did not utilize the thermal treatment described above which appears essential to achieve maximum conversion into the desired isomer.

5. Any commercially available autoclave, such as those used for high pressure hydrogenations, appears to be satisfactory.
6. The substance can readily be examined by VPC, using either an SE-30 or Carbowax-20M column.
7. This in another example of a relatively difficult oxidation (**P6.10**).
8. For the preparation of *cis,cis,trans*-perhydro-9*b*-phenalenol, it is not necessary to isolate either *cis,trans*-perhydro-9*b*-boraphenalene or *cis,cis, trans*-2-(perhydro-9′*b*-phenalyl)-1,3,2-dioxaborole. The thermally treated mixture containing the cyclic organoborane can be directly carbonylated without isolation, and the carbonylated crude product can be oxidized directly. In this way, a 92:8 mixture of *cis,cis,trans*- and *cis,cis,cis*-perhydro-9*b*-phenalenol is obtained in a yield of 70 to 75%.
9. A similar procedure can be employed for the preparation of the all-*cis* isomer.[3]

[1] H. C. Brown and E. Negishi, *J. Amer. Chem. Soc.*, **89**, 5478 (1967).
[2] G. W. Rotermund and R. Köster, *Ann.*, **686**, 153 (1965).
[3] H. C. Brown and W. C. Dickason, *J. Amer. Chem. Soc.*, **91**, 1226 (1969); W. C. Dickason, Ph.D. thesis, Purdue University, 1970.

8.8. *CIS,ANTI,TRANS*-9-KETOPERHYDROPHENANTHRENE

$$+ \text{BH}_2 \xrightarrow[0^\circ]{\text{THF}}$$

$$+ \text{KCN} + (\text{CF}_3\text{CO})_2\text{O} \xrightarrow[-78^\circ \text{ to } 25^\circ]{\text{THF}}$$

B, N, O, C, CF_3

$+ KO_2CCF_3$

$$+ 2\,H_2O_2 + 2\,NaOH + H_2O \xrightarrow[25^\circ]{THF}$$

$$+ NH_3 + NaO_2CCF_3 + \text{—OH}$$

Procedure by Andrew Pelter and D. J. Williams[1,2]

Procedure

The oven-dried apparatus consists of a 100-ml three-neck flask, one neck of which is connected to a stopcock bearing a septum cap, another neck is connected through a stopcock to a nitrogen line or to a vacuum pump, and the third contains a sealed tube into which the potassium cyanide is weighed prior to reaction. The inclination of the tube is such that by rotating it in the socket the solid will fall cleanly into the reaction mixture **(S9.5)**. In the sidearm is placed 0.7 g of dry, finely divided potassium cyanide (~8% excess), and in the flask is placed a magnetic stirring bar. The apparatus is evacuated and then nitrogen is introduced and a static pressure of nitrogen is maintained until following the oxidation. Thexylborane **(P2.10)**, 10 mmoles in 30 ml of tetrahydrofuran **(S9.11)** is placed in one syringe of a pair which operates mechanically to deliver at equal rates. In the other syringe is placed 1.76 g (10 mmoles) of 1-(2′-methylenecyclohexyl)cyclohexene **(N1)** in 30 ml of tetrahydrofuran. In the flask is placed 10 ml of tetrahydrofuran. The reaction flask is immersed in an ice-water bath, and to the stirred solvent in the flask, the two solutions in the syringes are added simultaneously over 3 hours at equal rates **(N2)** through needles inserted through the serum cap. The solution is then maintained at 25° for 1 hour to complete the reaction. The potassium cyanide is added by rotating the sidearm, and the reaction mixture is stirred for 1 hour, after which the solid has largely dissolved to form the cyanide complex. The flask is cooled to −78°, and 1.68 ml (12 mmoles) of trifluoroacetic anhydride is added through the septum cap. The cooling bath is then removed, and the reaction mixture is allowed to stir for 1 hour as it comes to room temperature. To the reaction mixture is rapidly added 10 ml of 5*M* sodium hydroxide, followed by the slow dropwise addition of 10 ml of 50% hydrogen peroxide (exothermic!) **(N4)**. The reaction mixture is allowed to stand overnight **(N3)**. The solution is filtered, and the pentane and thexyl alcohol are removed at the pump. The crude product in pentane is put on a dry silica column. Elution with pentane gives some starting olefin (0.135 g, 0.77 mmole), and elution with methylene chloride gives the crystalline ketone, *cis,anti,trans*-9-ketoperhydrophenanthrene, mp 60-62° (1.52 g. 80% yield, 92% conversion).

Notes

1. Condensation of cyclohexanone under the influence of pyrophosphoric acid provides 2-(1-cyclohexenyl)cyclohexanone, bp 108° (2 mm), n^{20}D 1.4999.[3] Treatment with methylenetriphenylphosphorane[4] provides 1-(2′-methylene-cyclohexyl)cyclohexene, bp 59-61° (0.3 mm), n^{23}D 1.5016 (**S9.11**).
2. A Sage syringe pump or equivalent is satisfactory. Alternatively, the two solutions can be added from a pair of dropping funnels although this is not the recommended procedure. The procedure followed for the cyclic hydroboration of a diene with thexylborane is essentially that described in **P4.16** for limonene.
3. The oxidation is generally complete in 3-4 hours, but it was convenient in this case to leave it overnight.
4. In related reactions 30% hydrogen peroxide has proved effective.

[1] Unpublished results by Andrew Pelter and D. J. Williams, University College of Swansea, Swansea SA2 8PP, Wales, UK.
[2] A. Pelter, M. G. Hutchings, and K. Smith, *Chem. Commun.*, 1048 (1971).
[3] K. U. Kelly and J. S. Matthews, *J. Chem. Eng. Data,* **14**, 277 (1969).
[4] G. Wittig and U. Schöllkopf, *Org. Syn.*, **40**, 66 (1960).

8.9. THEXYLCYCLOHEXYLCYCLOPENTYLCARBINOL

B + $HCCl_2OCH_3$ + $LiOCEt_3$ $\xrightarrow[0°]{THF}$ C–B(Cl)(OCH_3) + Et_3COH + LiCl

C–B(Cl)(OCH_3) + $(CH_2OH)_2$ $\xrightarrow{25°}$ C–B(O–CH_2)(O–CH_2) + HCl + CH_3OH

$$\text{Thexyl(cyclopentyl)(cyclohexyl)C-B(OCH_2CH_2O)} + H_2O_2 + NaOH + 2\,H_2O \xrightarrow[50^\circ]{\text{THF-EtOH}} \text{Thexyl(cyclopentyl)(cyclohexyl)C-OH} + NaB(OH)_4 + (CH_2OH)_2$$

Procedure by Jean-Jacques Katz and Bruce A. Carlson[1]

Procedure

A 500-ml flask fitted with a septum inlet, magnetic stirring bar, pressure-equalizing funnel, reflux condenser, and mercury bubbler is assembled, flushed with nitrogen, and maintained under a static pressure of the gas (**S9.1**) until the reaction with α,α-dichloromethyl methyl ether is complete. In the flask is prepared 0.100 mole of thexylcyclohexylcyclopentylborane (**P4.18**). The solution is brought to 0°, and 25.3 g (0.22 mole, 120% excess) of α,α-dichloromethyl methyl ether (**N1**) is added (**P8.10**:**N3**). In the dropping funnel is placed 111 ml of a 1.80*M* solution of lithium triethylcarboxide (0.200 mole, 100% excess) (**N2**). The base is added to the reaction mixture at 0° over 20-30 minutes, and the mixture is then maintained at room temperature for 30 minutes. A heavy precipitate of lithium chloride forms. To the reaction mixture is added 12.4 g (0.200 mole, 100% excess) of ethylene glycol and the volatile materials removed by distillation (**N3**). The product, 2-[2′,3′-dimethyl-2′-butyl)cyclohexylcyclopentylcarbinyl]-1,3,2-dioxaborole can be isolated by distillation, as described for the related perhydro-9′*b*-phenalyl derivative (**P8.7**). However, the more usual procedure is the direct oxidation to the corresponding carbinol (**P6.10**) to provide a synthetic route to highly hindered tertiary alcohols.[1]

Notes

1. The commercially available reagent (Aldrich) was distilled and stored under nitrogen. Use of the commercial material without distillation resulted in low yields.
2. Lithium triethylcarboxide in hexane is conveniently prepared by the dropwise addition of one equivalent of triethylcarbinol to one equivalent of *n*-butyllithium at 0° under nitrogen.
3. The oxidation is facilitated by the formation of the cyclic ester (**P6.10**:**N2**).

[1] H. C. Brown, J.-J. Katz, and B. A. Carlson, *J. Org. Chem.*, **38**, 3968 (1973).

8.10. BICYCLO[3.3.1]NONAN-9-ONE

H B + H_3C OH CH_3 $\xrightarrow[25°]{THF}$ H_3C CH_3 O B

H_3C CH_3 O B + $HCCl_2OCH_3$ + 2 $LiOCEt_3$ $\xrightarrow[0°]{THF}$ CH_3 $OCEt_3$ O–B OCH_3 C CH_3

+ Et_3COH + 2 LiCl

CH_3 $OCEt_3$ O–B OCH_3 C CH_3 + NaOH + H_2O_2 $\xrightarrow[50°]{THF, EtOH}$ O C

+ CH_3OH + H_3C OH CH_3 + Et_3COH + $NaB(OH)_4$

Procedure by Bruce A. Carlson[1]

Procedure

A 1-ℓ dry flask fitted with a septum inlet, reflux condenser, and magnetic stirring bar is attached to a mercury bubbler. The system is flushed with nitrogen. The flask is removed, and 12.2 g (0.100 mole) of solid 9-BBN (**P2.11** or Aldrich) is introduced with minimum exposure to the atmosphere (**N1**). The flask is reattached to the system, which is again flushed with nitrogen and maintained under a constant pressure of the gas (**S9.1**) until following the oxidation. Into the flask is introduced 50 ml of dry tetrahydrofuran (**S9.11**). Then to the reaction mixture is added 12.2 g (0.100 mole) of 2,6-dimethylphenol in 10 ml of tetrahydrofuran. The evolution of hydrogen is complete within 2 hours at room temperature (**N2**). The flask is then immersed in an ice-water bath and the solution brought to approximately 0°. To the cooled solution is added 10 ml, 12.6 g (0.110 mole, 10% excess) of α,α-dichloromethyl methyl ether (**N3**), followed by the slow addition over 20 minutes of 125 ml of a 1.60*M* solution of lithium triethylcarboxide (0.200 mole) in hexane (**N4**). The reaction is slightly exothermic and a white solid, presumably lithium chloride, precipitates. Following the addition, the reaction mixture is stirred for 0.5 hour at room temperature. To the reaction mixture is added 80 ml of 95% ethanol, 20 ml of water, and 12.0 g (0.300 mole) of solid sodium hydroxide. The flask is immersed in an ice bath and oxidation carried out by the slow, dropwise addition of 30 ml of 30% hydrogen peroxide (exothermic reaction! **P2.1:N6**), maintaining the temperature below 50°. After the initial vigorous reaction has subsided, the mixture is heated to maintain the temperature at 50° for 1 hour to ensure completion of the oxidation. The aqueous phase is saturated with sodium chloride. The organic phase is removed, washed once with 50 ml of a saturated salt solution, and the solvents removed on a rotary evaporator. The orange liquid is diluted with 100 ml of pentane and the phenol extracted with 3*M* aqueous sodium hydroxide (1 × 50 ml, and 1 × 30 ml). The pentane solution is washed with 50 ml of saturated salt solution. The pentane is removed under reduced pressure, followed by triethylcarbinol, bp 54-56° (16 mm). The semi-solid residue is dissolved in 75 ml of pentane and the solution cooled to −78° to crystallize the product. The product is collected on a filter and washed with cold (−78°) pentane (2 × 10 ml). The first crop of crystals provides 11.8 g of bicyclo[3.3.1]nonan-9-one, mp 154-156.5°, DNP, mp 190.5-191.2° [lit.[2] mp 155-158.5°, DNP, mp 191.8-192.3°]. Concentration of the mother liquor, followed by crystallization at −78°, yields a second crop of 0.3 g, mp 153-155°, a total isolated yield of 88%.

Notes

1. Alternatively, the 9-BBN can be placed in a small flask attached to the reaction flask by wide flexible tubing or by a rotating assembly utilized for the introduction of solids under a nitrogen atmosphere (**S9.5**).
2. Approximately 2.5 ℓ of hydrogen (25°) is evolved. The gas should be conducted away and safely vented.
3. Unlike chloromethyl methyl ether and 1,1′-dichloromethyl methyl ether, α,α-dichloromethyl methyl ether has been found to have no significant carcinogenic activity.[3]
4. Prepared by adding triethylcarbinol dropwise to *n*-butyllithium in hexane at 0° under nitrogen.

[1] B. A. Carlson and H. C. Brown, *Synthesis,* 776 (1973).
[2] C. F. Foote and R. B. Woodward, *Tetrahedron,* **20**, 687 (1963).

[3]B. L. Van Duuren, C. Katz, B. M. Goldschmidt, K. Frenkel, and A. Sivak, *J. Nat. Cancer Inst., USA,* **48,** 1431 (1972).

8.11. ETHYL(*TRANS*-2-METHYLCYCLOPENTYL)ACETATE

CH_3 (1-methylcyclopentene) + 9-BBN (BH) $\xrightarrow[67°]{THF}$ B-(*trans*-2-methylcyclopentyl)-9-BBN

B-(*trans*-2-methylcyclopentyl)-9-BBN + $CH_2BrCO_2C_2H_5$ + *t*-BuOK $\xrightarrow{0°}$ *trans*-2-methylcyclopentyl-$CH_2CO_2C_2H_5$ + KBr + *t*-BuOB(9-BBN)

Procedure by Milorad M. Rogić[1]

Procedure

A dry 500-ml flask equipped with a septum inlet, dropping funnel, magnetic stirring bar, and mercury bubbler was flushed with nitrogen and maintained under a static pressure of the gas **(S9.1)** until after the oxidation. 1-Methylcyclopentene **(S9.11)** is hydroborated with 9-BBN in tetrahydrofuran under reflux to produce 0.100 mole of B-(*trans*-2-methylcyclopentyl)-9-BBN **(P4.4)**. To the stirred solution at 0° is added 50 ml of dry *t*-butyl alcohol and 16.7 g (0.100 mole) of ethyl bromoacetate. Reaction is induced by the slow dropwise addition of 100 ml of a 1.00*M* solution of potassium *t*-butoxide in *t*-butyl alcohol over a period of 30 minutes **(N1)**. The reaction mixture is brought to room temperature. To the mixture is added 33 ml of 3*M* sodium acetate, followed by the dropwise addition of 22 ml of 30% of hydrogen peroxide (exothermic! **P2.1:N6) (N2)**, maintaining the temperature between 40-50°. The reaction mixture is stirred at room temperature for 30 minutes, and then the aqueous phase is saturated with sodium chloride. The organic layer is separated, dried over anhydrous magnesium sulfate, and distilled. There is obtained 9.7 g (57% yield) of ethyl (*trans*-2-methylcyclopentyl)acetate, bp 80° (10 mm), n^{20}D 1.4383 **(N3, N4)**.

Notes

1. Tris(*trans*-2-methylcyclopentyl)borane fails to undergo this reaction.[1] Apparently, the boron atom is too hindered.

2. Oxidation is for the removal of the B–OH–9-BBN residue.
3. VPC examination reveals the product to be at least 98% pure *trans*. The product was converted to *trans*-1-methyl-2-*n*-propylcyclopentane and compared with an authentic sample.
4. Alternatively, B-*p*-tolyl-9-BBN (**P4.7**) can be used to arylate phenacyl bromide by the same procedure to give the product in a yield of 95%.[2]

$$C_6H_5COCH_2Br + \text{B-}p\text{-tolyl-9-BBN} \xrightarrow{t\text{-BuOK}} C_6H_5COCH_2C_6H_4CH_3 \quad 95\%$$

Ethyl bromoacetate is converted into ethyl *p*-tolylacetate in a yield of 73%.

[1]H. C. Brown, M. M. Rogić, M. W. Rathke, and G. W. Kabalka, *J. Amer. Chem. Soc.*, **91**, 2150 (1969).
[2]H. C. Brown and M. M. Rogić, *J. Amer. Chem. Soc.*, **91**, 4304 (1969).

8.12. ETHYL 3-OCTENOATE

$$(n\text{-}C_4H_9)_3B + CH_2BrCH{=}CHCO_2C_2H_5 + \text{2,6-di-}t\text{-butylphenoxide } O^-K^+ \xrightarrow[\text{THF}]{0°}$$

$$n\text{-}C_4H_9CH_2{=}CHCH_2CO_2C_2H_5 + n\text{-}Bu_2BOR + KBr$$

Procedure by Hirohiko Nambu[1]

Procedure

A dry 200-ml flask, equipped with a septum inlet, dropping funnel, condenser, magnetic stirring bar, and mercury bubbler, is flushed with nitrogen and maintained under a constant pressure of nitrogen (**S9.1**). The flask is charged with 11.4 g (0.055 mole) of pure 2,6-di-*t*-butylphenol in 30 ml of tetrahydrofuran (**S9.11**), followed by 50 ml of 1.00*M* potassium *t*-butoxide in tetrahydrofuran.

The flask is immersed in an ice bath, and 9.1 g (12.2 ml, 0.050 mole) of tri-*n*-butylborane (**N1**) is added, followed by the dropwise addition over 30 minutes of 9.65 g (0.050 mole) of ethyl 4-bromocrotonate (**N2**) in 20 ml of tetrahydrofuran. The reaction mixture is allowed to stir for 1 hour at 0°. VPC analysis indicates a 72% yield of ethyl 3-octenoate. The temperature is brought to 25° and the residual organoboron intermediates oxidized by adding 16.5 ml of 3*M* sodium acetate followed by 12 ml of 30% hydrogen peroxide (exothermic! **P2.1:N1**) at a rate sufficient to maintain the temperature below 35°. After stirring for an additional 30 minutes, sodium chloride is added to saturate the aqueous phase. The organic phase is separated, dried over magnesium sulfate, and distilled. There is obtained 5.1 g (60% yield) of ethyl 3-octenoate, bp 93-95° (10 mm), n^{20}D 1.4362 (**N3**).

Notes

1. Tri-*n*-butylborane can be prepared following the procedure described for tri-*sec*-butylborane (**P2.7**). Alternatively, tri-*n*-butylborane is available commercially (Callery) (d^{25} = 0.747), as well as 1.0*M* solutions of tri-*n*-butylborane in tetrahydrofuran (Aldrich).
2. Commercial, bp 47-48° (0.4 mm), n^{20}D 1.4911 (**S9.11**).
3. VPC examination indicated the presence of only one isomer, and the IR spectrum corresponded to the presence of the *trans* isomer. However, VPC examination of the alcohols, obtained by reducing the esters with lithium aluminum hydride, revealed the presence of a small amount (~20%) of the *cis* isomer. The reaction provides a convenient means of achieving a 4-carbon homologation.

[1] H. C. Brown and H. Nambu, *J. Amer. Chem. Soc.*, **92**, 1761 (1970).

8.13. 1-(10-CARBOMETHOXYDECYL)CYCLOPENTANOL

$$\text{ThxBH}_2 + \text{cyclopentene} \xrightarrow[-20^\circ]{\text{THF}} \text{Thx(cyclopentyl)BH}$$

$$\text{Thx(cyclopentyl)BH} + CH_2{=}CH(CH_2)_8CO_2CH_3 \xrightarrow[0^\circ]{\text{THF}} \text{Thx(cyclopentyl)B}(CH_2)_{10}CO_2CH_3$$

$(CH_2)_{10}CO_2CH_3$ + Br_2 + H_2O $\xrightarrow[0°, h\nu]{CH_2Cl_2}$ OH $(CH_2)_{10}CO_2CH_3$ + 2 HBr

OH $(CH_2)_{10}CO_2CH_3$ + NaOAc + H_2O_2 $\xrightarrow[40°]{C_2H_5OH}$ HO $(CH_2)_{10}CO_2CH_3$

Procedure by Yoshinori Yamamoto and Clinton F. Lane[1]

Procedure

A dry 500-ml flask equipped with a septum inlet, pressure-equalizing dropping funnel, reflux condenser, magnetic stirring bar, and a mercury bubbler is flushed with nitrogen and then maintained under a static pressure of the gas **(S9.1)** until following the oxidation step. The flask is immersed in ice-water bath, and 100 ml of 1.00*M* thexylborane in tetrahydrofuran is prepared as described in **P2.10**. To this solution is added 6.8 g (0.100 mole) of cyclopentene **(S9.11)**, and stirring at −20° is continued for 1 hour. Then 19.8 g (0.100 mole) of methyl 10-undecenoate **(P6.7:N1)** is added and the reaction mixture stirred for another hour at 0°. The apparatus is connected to an aspirator, and the tetrahydrofuran is removed under reduced pressure. Nitrogen is used to restore the system to atmospheric pressure. The flask is charged with 70 ml of methylene chloride and 50 ml of water. The flask is again immersed in an ice-water bath, brought to approximately 0°, and 16.0 g (0.100 mole) of bromine in 30 ml of methylene chloride is slowly added to the flask as it is illuminated by a 200-W sunlamp **(N1)**. The reaction is permitted to proceed at 0° until the bromine color disappears (approximately 1 hour). Aqueous sodium hydroxide, 36 ml of 6*M*, is added to neutralize the hydrogen bromide produced. Oxidation of the intermediate is accomplished by adding 100 ml of ethanol, 33 ml of 3*M* aqueous sodium acetate, followed by the careful dropwise addition of 33 ml of 30% hydrogen peroxide (exothermic! **P2.1:N6**), maintaining the temperature below 50°. After 1 hour at 50°, the reaction mixture is cooled, the aqueous phase is saturated with sodium chloride, and the organic phase is separated. The aqueous phase is extracted once with 50 ml of methylene chloride. The combined organic layer is washed first with 50 ml of water saturated with sodium bicarbonate, and then with 50 ml of water saturated with sodium chloride. After drying the organic phase over anhydrous magnesium sulfate, distillation yields 20.7 g (73% yield) of 1-(10-carbomethoxydecyl)cyclopentanol, bp 110-112° (0.001 mm), n^{20}D 1.4619 (N2, N3).

Notes

1. In most cases, the thexyldialkylboranes are permitted to react at room temperature without the special illumination provided by the sunlamp. In the present case the reaction temperature was lowered to 0° and the rate enhanced by use of the sunlamp in order to minimize hydrolysis of the ester group under the influence of the hydrobromic acid produced in the course of the reaction. Bromine may be handled using glass-tipped syringes and Teflon needles.
2. This alcohol is prone to undergo dehydration during distillation at higher temperatures.
3. The related reaction involving cyclopentene and ethyl 3-butenoate reveals interesting differences.[1] In this reaction, a total of 3 moles of bromine per mole of borane are required to achieve optimum yields, 90% of 1-(3-carbomethoxypropyl)cyclopentanol. Apparently, coordination of the carbonyl group of the ester with the boron atom reduces the activity of the α-position. Attack then occurs competitively at both this position and the tertiary position of the thexyl group. The latter derivative eliminates 2,3-dimethyl-2-butene, which appears in the product as the dibromide.

[1] H. C. Brown, Y. Yamamoto, and C. F. Lane, *Synthesis,* 304 (1972).

8.14. BICYCLO[3.3.0]OCTAN-1-OL

H–B(bicyclic) + CH_3OH $\xrightarrow[\text{THF}]{25^\circ}$ OCH_3–B(bicyclic) + H_2

OCH_3–B(bicyclic) + CH_3Li $\xrightarrow[-78^\circ \rightarrow 25^\circ]{\text{pentane}}$ CH_3–B(bicyclic) + $LiOCH_3$ ↓

CH_3–B(bicyclic) + Br_2 $\xrightarrow[h\nu,\ 0^\circ]{CH_2Cl_2}$ CH_3–B(bicyclic)–Br + HBr

$$\text{B-methyl-9-bromo-9-BBN} + H_2O \xrightarrow{0^\circ} \text{CH}_3\text{B(OH)-bicyclo[3.3.0]octane} + HBr$$

$$\text{CH}_3\text{B(OH)-bicyclo[3.3.0]octane} + H_2O_2 + NaOH \longrightarrow \text{bicyclo[3.3.0]octan-1-ol} + NaB(OH)_4$$

Procedure by Norman R. De Lue and Gary W. Kramer[1,5]

Procedure

This procedure involves (a) the preparation of B-methoxy-9-BBN; (b) the preparation of B-methyl-9-BBN; and (c) the bromination-hydrolysis of B-methyl-9-BBN. We have chosen to describe each ot these steps separately and have included isolation procedures for B-methoxy-9-BBN and B-methyl-9-BBN. However, the preparation of the bicyclo[3.3.0]octane-1-ol can be carried out from 9-BBN in one pot without isolation of the methoxy or methyl derivatives. The overall yield using this *in situ* procedure is 61% based on 9-BBN **(N1)**.

a. B-Methoxy-9-BBN.[1,2] An oven-dried 500-ml flask fitted with a septum inlet, magnetic sitrring bar, and splash guard is connected to a mercury bubbler and flushed thoroughly with nitrogen. The flask is removed and charged, in a nitrogen-filled glove bag, with 25 g of solid 9-BBN (0.205 mole) **(P2.11** or Aldrich). The flask is reconnected, and the entire apparatus is purged with nitrogen and maintained under a positive pressure of the gas throughout the remainder of the experiment. About 50 ml of dry tetrahydrofuran **(S9.11)** is added to get a stirrable slurry. Then 30 ml of absolute methanol is added dropwise from a syringe using a mechanical syringe pump **(N2, N3)**. After stirring for an additional hour, the solvent and excess methanol are removed by distillation at atmospheric pressure. The residual oil is then distilled under vacuum **(S9.7)** to give 27 g (87%) of B-methoxy-9-BBN, bp 30-33° (0.08 mm). This material was 99% pure by VPC and ^{1}H NMR. ^{1}H NMR (CCl_4): 1.28 δ, 1.80 δ (multiplet, 14H, rings), and 3.70 δ (s, 3H, OCH_3). ^{11}B NMR (cyclopentane): -52 δ; n^{20}D 1.4792.

b. B-Methyl-9-BBN.[1] An oven-dried 300-ml flask fitted with a septum inlet and a magnetic stirring bar is connected to a mercury bubbler, flushed with nitrogen, and maintained under a positive pressure of the gas throughout the preparation. The flask is charged with 22.8 g of B-methoxy-9-BBN (0.15 mole) and 100 ml of dry, olefin-free pentane **(S9.11)**. Stirring is begun, and the flask is cooled in a Dry Ice-acetone bath. Standardized methyllithium in ether (75 ml of 2.00*M*,

0.15 mole) **(S9.10)** is added slowly from a graduated cylinder via double-ended needle **(S9.2, N4)**. After stirring about 10 minutes at -78°, allow the reaction mixture to warm to room temperature and stir for 3 hours or more. The solid lithium methoxide is allowed to settle, and the supernatant liquid is decanted via double-ended needle into an evacuated distillation apparatus where the solvent is flashed off **(S9.7)**. The solid is washed with 2 X 75 ml of pentane and allowed to settle, and the supernatant liquid is decanted into the distillation assembly as above **(N5)**. The residual oil is vacuum distilled to give 18 g (88%) of B-methyl-9-BBN (bp 23-24° at 0.5 mm; bp 64-65° at 15 mm). The product is greater than 99% pure by VPC and ^{1}H NMR. ^{1}H NMR (neat): 1.2 δ, 1.8 δ (multiplet, 14H, rings), and 0.92 δ (s, 3H, CH_3). ^{11}B NMR (cyclopentane): -86 δ.

c. Bicyclo[3.3.0]octan-1-ol.[5] A 1-l flask, equipped with a septum inlet, thermometer well, pressure-equalized addition funnel, reflux condenser, and magnetic stirring bar, is flushed with dry nitrogen and maintained under a positive nitrogen pressure of the gas until after the oxidation **(S9.1)**. The flask is charged with 13.6 g of B-methyl-9-BBN (15.8 ml, 0.100 mole), 100 ml of methylene chloride, and 200 ml of water. The flask is then cooled to 0-5°, and 5.15 ml of bromine in 100 ml of methylene chloride is added as such a rate that the reaction mixture never becomes more than light yellow in color (1-2 hours). Vigorous stirring is maintained during the bromination. After the bromine color has disappeared, there is added 200 ml of methanol and 100 ml of 6*M* aqueous sodium hydroxide. Oxidation is carried out by the slow addition of 40 ml of 30% aqueous hydrogen peroxide followed by heating at reflux for 1 hour **(N6)**. The reaction mixture is cooled, and the aqueous phase is saturated with potassium carbonate. The organic layer is separated, washed with saturated aqueous sodium chloride (1 X 50 ml), and dried over anhydrous potassium carbonate. Volatiles are removed by rotary evaporator, and the residual oil is vacuum distilled to give 11.1 g (88% yield) of a waxy solid which solidifies in the receiver, bp 92-94° (16 mm). The *cis*-bicyclo[3.3.0]octan-1-ol is greater than 99% pure by VPC, *p*-nitrobenzoate, mp 124.0-124.5° (lit.[4] mp 124.0-124.8°) **(N7)**.

Notes

1. The apparatus described in (c) is charged with 0.100 mole of solid 9-BBN. After methanolysis, as described in (a), the volatiles are completely removed under aspirator vacuum and replaced with 100 ml of dry pentane. The crude B-methoxy-9-BBN is used to prepare B-methyl-9-BBN, as described in (b), Then 150 ml of water is added to dissolve the precipitated lithium methoxide. Five drops of 1% phenophthalein is added, and the solution is neutralized with 2*M* aqueous HCl until the pink color is discharged (~50 ml). The pentane and ether are removed under aspirator vacuum and replaced with 100 ml

of methylene chloride. Bromination and oxidation are carried out as described in (c). After work-up, distillation affords 7.7 g (61% yield, 95% pure by VPC of *cis*-bicyclo[3.3.0] octan-1-ol.

2. Spectral grade methanol dried over 3 A Molecular Sieves is adequate. The use of a syringe pump is convenient, but not necessary. The methanol can be added using a pressure-equalized addition funnel on top of the splash guard or very slowly by hand using a syringe.
3. Approximately 5 l of hydrogen is evolved and should be safely vented. Considerable foaming often occurs during the methanolysis.
4. Excess methyllithium should be avoided, since this results in lower yields, presumable due to formation of an insoluble "ate" complex.[1]
5. The washing process can be hastened by transferring the entire mixture (as a slurry) through a large bore double-ended needle into dry, nitrogen-flushed, 50-ml centrifuge tubes capped with rubber septa. The solid can be removed from the wash liquid by centrifuging.
6. This is another example of a difficult oxidation requiring ethanol cosolvent along with excess alkali and hydrogen peroxide **(P6.10)**.
7. Bromination of B-methoxy-9-BBN followed by oxidation provides *cis*-bicyclo-[3.3.0] octan-1-ol. However, the yield is lower and the reaction mixture is more complex.[3]

[1] G. W. Kramer and H. C. Brown, *J. Organometal. Chem.*, **73**, 1 (1974).
[2] E. F. Knights, Ph.D. Thesis, Purdue University, West Lafayette, Indiana, 1968.
[3] Y. Yamaoto and H. C. Brown, *J. Org. Chem.*, **39**, 861 (1974).
[4] R. C. Fort, Jr., R. E. Hornish, and G. A. Liang, *J. Amer. Chem. Soc.*, **92**, 7558 (1970).
[5] Research in progress with N. R. De Lue.

8.15. 2-ETHYLCYCLOHEPTANONE

$$\text{2-diazocycloheptanone} + (C_2H_5)_3B + H_2O \xrightarrow[25^\circ]{THF} \text{2-ethylcycloheptanone} + (C_2H_5)_2BOH + N_2$$

Procedure by John Hooz, Donald M. Gunn, and Hiromichi Kono[1]

Procedure

A 100-ml three-neck flask fitted with a septum inlet, reflux condenser, magnetic stirring bar, and a mercury bubbler is assembled, flushed with nitrogen, and maintained under a static pressure of that gas **(S9.1)** until following the oxidation. In the flask is placed 20 ml of tetrahydrofuran, 2.94 g (30 mmoles) of triethylborane **(N1)**, and 3.6 ml of water (0.200 mole). To the stirred solution (25°) is added 2.76 g (20 mmoles) of 2-diazocycloheptanone

(N2) in 30 ml of tetrahydrofuran. Nitrogen is rapidly evolved and is complete in 1 hour. VPC analysis reveals a yield of 92% of 2-ethylcycloheptanone. The reaction mixture is treated with 15 ml of 3*M* aqueous sodium acetate, followed by the dropwise addition of 15 ml of 30% hydrogen peroxide (exothermic!, **P2.1:N6**), maintaining the temperature below 50°. The aqueous layer is saturated with sodium chloride. The organic layer is separated, dried over magnesium sulfate, and distilled to provide 2.0 g (71% yield) of 2-ethylcycloheptanone, bp 85° (18 mm), n^{25}D 1.4610.

Notes

1. Commercial, from a cylinder (**S9.9**). Alternatively, the 0.5*M* solution of lithium triethylborohydride in tetrahydrofuran can be used and treated with one equivalent of methanesulfonic acid to generate the triethylborane *in situ*. The procedure utilizes triethylborane in excess, so that the precise amount need not be carefully controlled.
2. 2-Diazocycloheptanone, bp 62° (0.4 mm) is prepared in 83% yield by the reaction of *p*-toluenesulfonyl azide with 2-hydroxymethylenecycloheptanone in the presence of triethylamine at −12 to −15°.[2] For the present procedure it is not necessary to isolate the diazo intermediate by distillation; the crude product obtained by removal of the methylene chloride solvent is adequate. The synthesis of 2-diazocycloalkanones is applicable to the synthesis of ring derivatives from 5- to 12-carbon atoms[2] and the replacement of the diazo group by an ethyl group has been described for 5- to 8-rings, as well as more complex structures, such as 2-diazo-6-methylcyclohexanone, 2-diazo-1-indanone, and 16-diazoestrone methyl ether.[1]

[1] J. Hooz, D. M. Gunn, and H. Kono, *Can. J. Chem.*, **49**, 2371 (1971).
[2] M. Regitz, J. Rüter, and A. Liedhegener, *Org. Syn.*, **51**, 86 (1971).

8.16. ETHYL CYCLOPENTYLACETATE

$$(\text{C}_5\text{H}_9)_2\text{BCl} + \text{N}_2\text{CHCO}_2\text{C}_2\text{H}_5 \xrightarrow[\text{-78°}]{\text{EE}} [\ \ ] + \text{N}_2$$

$$[\ \ ] + \text{CH}_3\text{OH} \xrightarrow[\text{-78°}]{\text{EE}} \text{C}_5\text{H}_9\text{CH}_2\text{CO}_2\text{C}_2\text{H}_5$$

Procedure by M. Mark Midland and Alan B. Levy[1]

Procedure

A dry 500-ml flask equipped with a septum inlet, a magnetic stirring bar, and a mercury bubbler is flushed with nitrogen and maintained under a static pressure of the gas until after the methanolysis step. In the flask 0.200 mole of cyclopentene (**S9.11**) is treated with 0.100 mole of chloroborane etherate to produce dicyclopentylchloroborane in ethyl ether (**P4.14**). The solution is cooled to −78°, and to the rapidly stirred solution 11.4 g (0.100 mole) of ethyl diazoacetate is added dropwise over 20 minutes. The mixture is stirred for an additional 15 minutes at −78°, and then treated at this temperature with 10 ml of methanol, followed by 10 ml of water (**N1**). The reaction mixture is now allowed to come to room temperature. The aqueous layer is carefully saturated with potassium carbonate. The organic phase is separated and dried over magnesium sulfate. VPC analysis indicates a yield of 94%. Distillation yields 12.2 g (78% yield) of ethyl cyclopentylacetate, bp 98-100° (28 mm), n^{20}D 1.4398 (**N2**).

Notes

1. If the hydrolysis is deferred until the reaction mixture has come to room temperature, the yield of the desired product becomes negligible, approximately 7%. It was established that the intermediate is stable at −78° for at least 6 hours, but begins to be lost relatively rapidly at −30 to −40°.
2. The speed of this reaction is enormous compared to the related Hooz reaction involving the trialkylboranes.[2] This speed is doubtless a consequence of the greater Lewis acid strength of R_2BCl versus R_3B. Even more important than the speed is the fact that the reaction readily accommodates even bulky alkyl groups.

$$(\text{norbornyl})_2BCl \xrightarrow{N_2CHCO_2C_2H_5} \text{norbornyl}-CH_2CO_2C_2H_5$$

92%

A minor disadvantage remaining is the utilization of only one of the two alkyl groups in R_2BCl.

[1] H. C. Brown, M. M. Midland, and A. B. Levy, *J. Amer. Chem. Soc.*, **94**, 3662 (1972).
[2] J. Hooz and S. Linke, *J. Amer. Chem. Soc.*, **90**, 6891 (1968).

8.17. *CIS*-1-CYCLOHEXYL-1-HEXENE

$$n\text{-}C_4H_9C\equiv CH + (C_6H_{11})_2BH \xrightarrow[0^\circ]{THF} (Z)\text{-}n\text{-}C_4H_9CH{=}CHB(C_6H_{11})_2$$

$$n\text{-}C_4H_9CH{=}CHB(C_6H_{11})_2 + NaOH + I_2 \xrightarrow[-10^\circ]{THF} cis\text{-}n\text{-}C_4H_9CH{=}CHC_6H_{11} + NaI + C_6H_{11}B(OH)_3Na$$

Procedure by George Zweifel, H. Arzoumanian, and C. C. Whitney[1]

Procedure

A dry 500-ml flask equipped with a septum inlet, thermometer, a pressure-equalizing dropping funnel, a magnetic stirring bar, and a mercury bubbler is assembled, flushed with nitrogen, and maintained under a static pressure of nitrogen until following the reaction with iodine (**S9.1**). In the flask is placed 50 ml of a 2.00*M* solution of the borane-tetrahydrofuran complex (**P2.3, P2.4:N3**). The flask is immersed in an ice-water bath, and 16.4 g (0.200 mole) of cyclohexene (**S9.11**) in 50 ml of tetrahydrofuran (**S9.11**) is added through the dropping funnel. The slurry of dicyclohexylborane (**N1**) and tetrahydrofuran is stirred at 0-5° for 1 hour. To the reaction mixture is added 8.2 g (0.100 mole) of 1-hexyne maintaining the temperature at 0-5°. The precipitate dissolves as the vinylboron intermediate forms. The reaction mixture is now brought to room temperature and allowed to stir for an additional 30 minutes. The reaction flask is immersed in an ice-salt bath and the temperature lowered to −10°. To this mixture is added 60 ml of 6*M* sodium hydroxide, followed by the dropwise addition of a solution of 25.4 g (0.100 mole) of iodine in 40 ml of tetrahydrofuran over a period of 15 minutes. The reaction mixture is brought to room temperature, and excess iodine is removed by adding a small quantity of aqueous sodium thiosulfate. The aqueous phase is saturated with sodium chloride, and the tetrahydrofuran phase is separated and dried over anhydrous magnesium sulfate. Distillation yields 12.4 g (75% yield) of *cis*-1-cyclohexyl-1-hexene, bp 44-45° (1 mm), n^{20}D 1.4586.

Notes

1. Alternatively, dicyclohexylborane can be prepared by the procedure of **P2.8**. However, since it is utilized in the presence of tetrahydrofuran, the present procedure offers advantages of simplicity.
2. Cyclohexylboronic acid is also formed. However, since it dehydrates into the slightly volatile tricyclohexylboroxine, it does not interfere with the distillation of the product. In some cases, oxidation of the intermediate with alkaline hydrogen peroxide may be desirable.

[1] G. Zweifel, H. Arzoumanian, and C. C. Whitney, *J. Amer. Chem. Soc.*, 89, 3652 (1967).

8.18. *TRANS*-1-CYCLOHEXYL-1-HEXENE

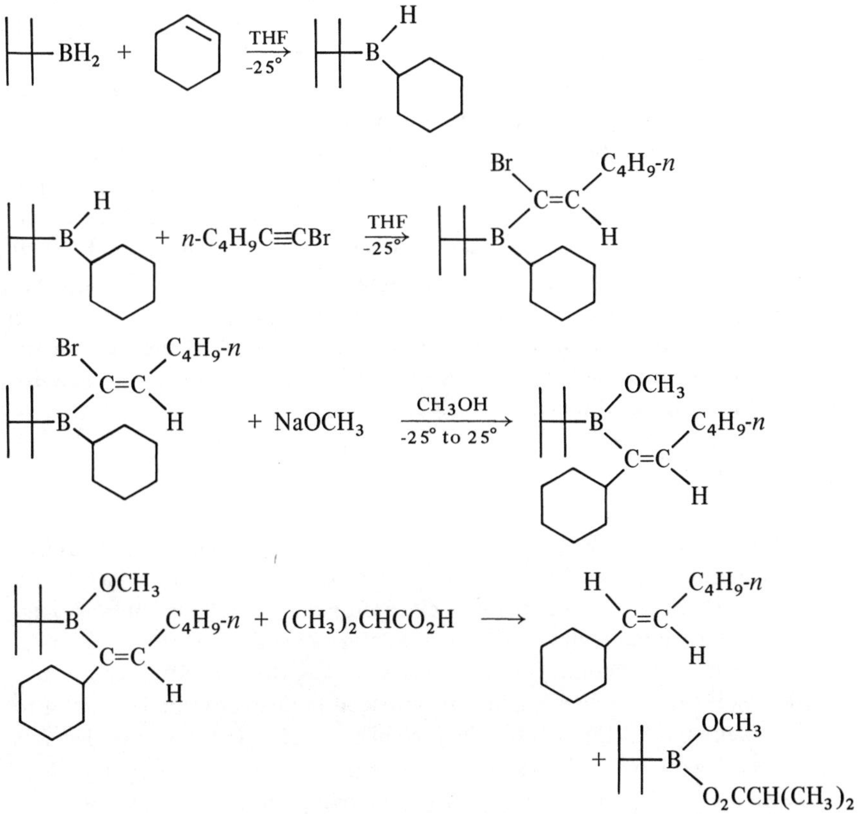

$-BH_2$ + (cyclohexene) $\xrightarrow[-25°]{THF}$ H, B

H, B + n-$C_4H_9C{\equiv}CBr$ $\xrightarrow[-25°]{THF}$ Br, C=C, C_4H_9-n, B, H

Br, C=C, C_4H_9-n, B, H + $NaOCH_3$ $\xrightarrow[-25° \text{ to } 25°]{CH_3OH}$ OCH_3, B, C=C, C_4H_9-n, H

OCH_3, B, C=C, C_4H_9-n, H + $(CH_3)_2CHCO_2H$ $\longrightarrow$ H, C=C, C_4H_9-n, H

+ B, OCH_3, $O_2CCH(CH_3)_2$

Procedure by Ei-ichi Negishi and Jean-Jacques Katz[1]

Procedure

A 500-ml three-neck flask equipped with a septum inlet, thermometer, pressure-equalizing dropping funnel, condenser, magnetic stirring bar, and mercury bubbler is assembled, flushed with nitrogen, and maintained under a positive pressure of nitrogen until following the protonolysis stage **(S9.1)**. In the flask is prepared 100 ml of a 1.00*M* solution of thexylborane in tetrahydrofuran **(P2.10)**. The solution is immersed in a cold bath, and the temperature is brought to −25°. To the flask is added 8.2 g (0.100 mole) of cyclohexene **(S9.11)**, and the reaction mixture is allowed to stir for 1 hour. Then 16.1 g (0.100 mole) of 1-bromo-1-hexyne **(N1)** is added and the mixture stirred for 1 hour at −25°. To the reaction mixture is slowly added 8.1 g (0.150 mole) of sodium methoxide in 100 ml of methanol (*Caution*: exothermic!) **(P6.7:N3)** **(N2)**. The reaction mixture is brought to room temperature and stirred for 1 hour, then 5 ml isobutyric acid is added to neutralize any excess sodium methoxide. The apparatus is attached to an aspirator and all of the volatile materials removed under vacuum (15 mm) at 25°. The pressure is briefly reduced to 1 mm to remove the last traces **(N3)**. Nitrogen is introduced to bring the system back to atmospheric pressure, and the condenser is again connected to the mercury bubbler. Then 100 ml of isobutyric acid is added to the reaction flask, and protonolysis is carried out by heating the reaction mixture under reflux for 1 hour. The cooled reaction mixture is poured into 300 ml of water and extracted with pentane (3 × 100 ml). The organic layer is washed with saturated, aqueous potassium carbonate (2 × 100 ml) to remove isobutyric acid and the pentane layer dried over magnesium sulfate. VPC analysis indicates a yield of 85% of the *trans* olefin, >99% pure. Distillation provides 13.3 g (80% yield) of *trans*-1-cyclohexyl-1-hexene, bp 44-45° (1.0 mm), n^{20}D 1.4569 **(N4)**.

Notes

1. 1-Bromo-1-hexyne is prepared by the method of Schulte and Goes.[2] Potassium hydroxide, 180 g (~3 moles), is placed in a 2-ℓ three-neck flask equipped with a mechanical stirrer and a dropping funnel. One liter of water is added to dissolve the alkali. The mixture is cooled to 0° and 27.5 ml (0.5 mole) of bromine is added dropwise (~30 minutes). Finally, 41 g (0.5 mole) of 1-hexyne is added dropwise at 0°. The mixture is allowed to stir overnight as the temperature comes to 25°. The aqueous layer is extracted with ether (3 × 100 ml). The ether layer is dried over anhydrous magnesium sulfate for 2 hours. After removing the solvent (25°, 30 mm), the product is distilled: 73.4 g (91% yield) of 1-bromo-1-hexyne, bp 44° (18 mm), n^{20}D 1.4657.

2. The procedure is essentially that of Zweifel,[3] modified to utilize the thexylmonoalkylboranes (**P4.17**) which offer the advantage that they are often more readily prepared than the dialkylboranes, except for a few special cases, and they avoid the loss of one of the alkyl groups.
3. Complete removal of the volatile components is desirable in order to maintain the temperature of protonolysis at, or near, the boiling point of isobutyric acid. If the refluxing temperature is less than 154°, a longer time is required to complete the protonolysis stage. These thexylvinylborinic esters undergo protonolysis much more slowly than representative dialkylvinylboranes (**S5.5, P6.5**).
4. It is possible to transfer groups such as *trans*-2-methylcyclopentyl with retention of stereochemistry.[1,4]

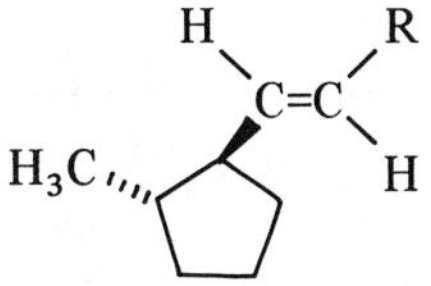

H R
C=C
H_3C H

[1] E. Negishi, J.-J. Katz, and H. C. Brown, *Synthesis,* 555 (1972).
[2] K. E. Schulte and M. Goes, *Anch. Pharm.,* **290,** 118 (1959).
[3] G. Zweifel and H. Arzoumanian, *J. Amer. Chem. Soc.,* **89,** 5086 (1967).
[4] G. Zweifel, R. P. Fisher, J. T. Snow, and C. C. Whitney, *J. Amer. Chem. Soc.,* **93,** 6309 (1971).

8.19. 4,5-DIETHYL-*CIS-TRANS*-3,5-OCTADIENE

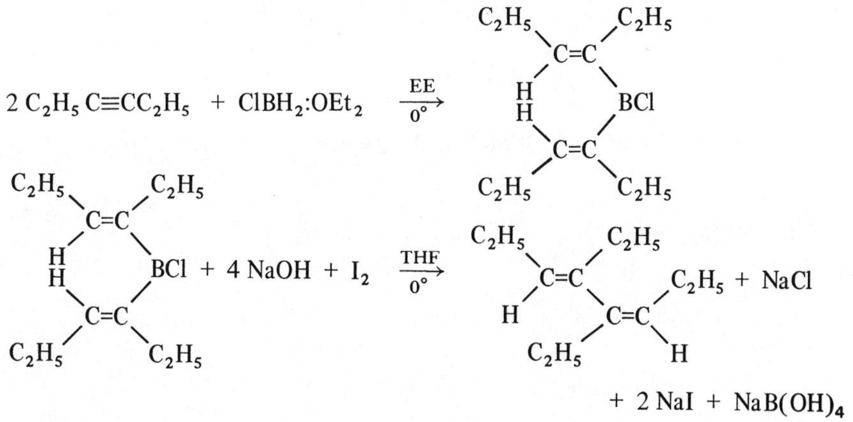

$$2\ C_2H_5C{\equiv}CC_2H_5 + ClBH_2{:}OEt_2 \xrightarrow[0^\circ]{EE} [(C_2H_5)C(H){=}C(C_2H_5)]_2BCl$$

$$[(C_2H_5)C(H){=}C(C_2H_5)]_2BCl + 4\ NaOH + I_2 \xrightarrow[0^\circ]{THF} (C_2H_5)(H)C{=}C(C_2H_5){-}C(C_2H_5){=}C(H)(C_2H_5) + NaCl$$

$$+ 2\ NaI + NaB(OH)_4$$

Procedure by Nair Ravindran[1]

Procedure

A 300-ml flask equipped with a septum inlet, magnetic stirring bar, and mercury bubbler is assembled. dried, flushed with nitrogen, and a static pressure of nitrogen maintained **(S9.1)** until following the treatment with iodine. In the flask is placed 8.2 g (0.100 mole) of 3-hexyne in 15 ml of anhydrous ethyl ether. The acetylene is converted into bis(*cis*-3-hexenyl)chloroborane **(P4.15)** by the addition via a hypodermic syringe of 50 ml of 1.00*M* chloroborane in ether **(P2.12)**. The reaction mixture is stirred for 2 hours at 0° to ensure completion of the hydroboration. The ether is removed under the vacuum of a water aspirator. The flask is immersed in an ice-water bath, and charged with 40 ml of tetrahydrofuran, followed by 67 ml of 3*M* aqueous sodium hydroxide (0.200 mole). The migration is accomplished by the dropwise addition of 12.7 g of iodine (0.050 mole) in 30 ml of tetrahydrofuran, until a slight color of residual iodine persists **(N1)**. The reaction mixture is brought to 25° and the excess iodine destroyed by stirring with excess sodium thiosulfate at 25°. The reaction mixture is extracted with pentane, washed with dilute sodium thiosulfate solution, dried, and distilled. There is obtained 6.9 g (83% yield) of 4,5-diethyl-*cis,trans*-3,5-octadiene, bp 62-64° (8 mm), n^{20}D 1.4542 **(N2)**.

Notes

1. The procedure is a modification of the Zweifel synthesis, avoiding the oxidation of initially formed thexyldialkenylboranes to produce the dialkenylborinic acids.[2]
2. The products appear to be isomerically pure by VPC and NMR examination.[1,2]

[1] H. C. Brown and N. Ravindran, *J. Org. Chem.*, **38**, 1617 (1973).
[2] G. Zweifel, N. L. Polston, and C. C. Whitney, *J. Amer. Chem. Soc.*, **90**, 6243 (1968).

8.20. 1-CYCLOHEXYL-*TRANS,TRANS*-1,3-OCTADIENE

$$\text{Thexyl}-BH_2 + ClC{\equiv}C\text{-}n\text{-}C_4H_9 \xrightarrow[-25^\circ]{\text{THF}} \text{Thexyl}-B(H)-C(Cl)=C(H)(n\text{-}C_4H_9)$$

$$\text{Thexyl}-B(H)-C(Cl)=C(H)(n\text{-}C_4H_9) + HC{\equiv}C\text{-}C_6H_{11} \xrightarrow[-25^\circ]{\text{THF}} \text{Thexyl}-B[C(H)=C(H)C_6H_{11}]-C(Cl)=C(H)(n\text{-}C_4H_9)$$

H C=C B H H + CH_3ONa $\xrightarrow[25°]{CH_3OH}$ C=C H H MeO C=C H B n-C_4H_9 + NaCl

Cl C=C n-C_4H_9

H C=C H MeO C=C H B n-C_4H_9 + $(CH_3)_2CHCO_2H$ $\xrightarrow{155°}$ H C=C H H C=C H n-C_4H_9

Procedure by Ei-ichi Negishi and Takao Yoshida[1]

Procedure

A dry 300-ml flask equipped with a septum inlet, a reflux condenser connected to a mercury bubbler, and a magnetic stirring bar is assembled, flushed with nitrogen, and maintained under a static nitrogen pressure until following the protonolysis **(S9.1)**. In the flask are placed 5.83 g (50 mmoles) of 1-chloro-1-hexyne **(N1)** and 40 ml of dry tetrahydrofuran **(S9.11)**. The solution is cooled to −25°, and 50 ml (50 mmoles) of 1.0*M* thexylborane in tetrahydrofuran **(P2.10)** is added slowly to the vigorously stirred solution. After stirring for 1 hour at −25°, 5.41 g (50 mmoles, 6.45 ml) of cyclohexylethyne is added dropwise at −25°. The reaction mixture is stirred for an additional hour at −25°. Then a solution of 4.05 g (75 mmoles) of sodium methoxide in 75 ml of methanol **(P6.7:N3)** is added at −25°. The cooling bath is removed after 5 minutes, and stirring is continued for 1 hour at room temperature. The mercury bubbler is disconnected, the reaction system connected to an aspirator, and the solvents and volatile materials are removed under a pressure of 15 mm at room temperature. The pressure is then reduced to 1 mm to remove the last traces of the solvents **(P8.18:N2)**. Nitrogen is introduced to bring the system back to atmospheric pressure, and the condenser is again connected to the mercury bubbler. Then 75 ml of isobutyric acid is added to the reaction flask, and protonolysis is carried out by heating the reaction mixture under reflux for 1 hour. The cooled reaction mixture is poured into 200 ml of water and extracted with diethyl ether. After washing the ether extract with a saturated aqueous solution of sodium carbonate to remove isobutyric acid, it is oxidized with 17 ml of 3*M* sodium hydroxide and 17 ml of 30% hydrogen peroxide to remove any boron-containing compounds (exothermic! **P2.1:N6**). The aqueous layer is extracted with diethyl ether, and the combined ether layer is washed with water and dried over magnesium sulfate. VPC analysis indicates a yield of 59% (isomeric purity, 98%). Distillation and column chromatography (Florisil, petroleum ether) provide 5.32 g (55% yield) of 1-cyclohexyl-*trans,trans*-1,3-octadiene, bp 112-115° (3 mm), n^{20}D 1.4905.

Notes

1. 1-Chloro-1-hexyne is prepared in a manner similar to that described in the literature.[2] The detailed procedure is as follows: a dry 500-ml flask equipped with a dropping funnel, a septum inlet, and an adaptor connected to a mercury bubbler is assembled and flushed with nitrogen. In the flask are placed 8.2 g (0.100 mole, 11.5 ml) of 1-hexyne and 75 ml of tetrahydrofuran. To this solution is added 52.5 ml (0.100 mole) of 1.9*M* *n*-butyllithium in hexane at −50°. After stirring for 1 hour at −50°, a solution of 19.1 g (100 mmoles) of *p*-toluene sulfonyl chloride in 40 ml of tetrahydrofuran is added from the dropping funnel at −50°. The reaction mixture is stirred for an additional 6 hours at −50° to −40°, and poured into 500 ml of water. The organic layer is extracted with pentane (total 200 ml). The pentane solution is washed with water and dried over magnesium sulfate. Distillation yields 7.5 g (64%) of 1-chloro-1-hexyne, bp 65-67° (125 mm).

[1] E. Negishi and T. Yoshida, *Chem. Commun.*, 606 (1973).
[2] H. Normant and T. Cuvigny, *Bull. Soc. Chim. France*, 1447 (1957); D. J. Pflaum and H. H. Wenzke, *J. Amer. Chem. Soc.*, **56**, 1106 (1934).

8.21. *TERT*-BUTYLPHENYLACETYLENE

$$(CH_3)_3CC{\equiv}CH + n\text{-BuLi} \xrightarrow[0^\circ]{THF} (CH_3)_3CC{\equiv}CLi + n\text{-BuH}$$

$$(CH_3)_3CC{\equiv}CLi + (C_6H_5)_3B \xrightarrow[0^\circ]{THF} [(CH_3)_3C{\equiv}CB(C_6H_5)_3]\,Li$$

$$[(CH_3)_3CC{\equiv}CB(C_6H_5)_3]\,Li + I_2 \xrightarrow[-78^\circ]{THF,\ EE} (CH_3)_3CC{\equiv}CC_6H_5 + (C_6H_5)_2BI + LiI$$

Procedure by M. Mark Midland et al.[1]

Procedure

A dry 500-ml flask equipped with a septum inlet, pressure-equalizing dropping funnel, magnetic stirring bar, and mercury bubbler is flushed with nitrogen and maintained under a static pressure of the gas until following the oxidation (**S9.1**). The flask is charged under nitrogen with 7.3 g (30 mmoles) of triphenylborane (**S9.5**) and 25 ml of dry tetrahydrofuran (**S9.11**). The flask is immersed in an ice-water bath and the contents brought to 0°. In a dry 100-ml flask maintained under nitrogen is placed 50 ml of tetrahydrofuran and 2.46 g (30.0 mmole) of 3,3-dimethyl-1-butyne (**N2**). The flask is cooled in an ice bath, and 16.0 ml (30.0 mmoles) of a 1.88*M* solution of *n*-butyllithium in hexane (**S9.10**) is added to form the lithium acetylide. The lithium acetylide solution is then transferred to the 500-ml flask with the aid of a double-ended needle (**S9.2**). The reaction mixture is stirred for several minutes to form the "ate" complex, and the solution is then cooled to −78°. A solution of 7.62 g (30 mmoles) of iodine in 75 ml of ethyl ether is then introduced into the dropping funnel, and the iodine solution is added dropwise to the stirred reaction mixture over 0.5 hour. After an additional 45 minutes at −78°, the reaction mixture is allowed to come to room temperature. The solution is then washed twice with 20 ml of 3*M* sodium hydroxide (containing 1 ml of a saturated aqueous solution of sodium thiosulfate to remove residual iodine). The combined aqueous phase is extracted with 25 ml of ethyl ether. The combined organic phase is treated with 32 ml of 3*M* sodium hydroxide (**N3**) followed by the dropwise addition of 10.5 ml of 30% hydrogen peroxide (exothermic! **P2.1:N6**) to oxidize the diphenylborinic acid by-product. Potassium carbonate is added to saturate the aqueous phase (**P2.11:N8**). VPC analysis of the organic phase reveals a 94% yield of product. The organic layer is separated, dried over potassium carbonate, and distilled. There is obtained 3.93 g (83% yield) of 1-phenyl-3,3-dimethyl-1-butyne, bp 100° (20 mm), n^{20}D 1.5175 [lit.[2] bp 84° (10 mm), n^{20}D 1.5230].

Notes

1. Triphenylborane, mp 147°, must be handled under nitrogen.
2. Bp 39-40°, n^{20}D 1.3749. The compound may be purified by distillation from a small amount of sodium borohydride (**S9.11**).
3. Three moles of base per mole of borinic acid are used here to provide adequate alkali to neutralize the two moles of phenol and the mole of boric acid formed in the oxidation.

[1] A. Suzuki, N. Miyaura, S. Abiko, M. Itoh, H. C. Brown, J. A. Sinclair, and M. M. Midland, *J. Amer. Chem. Soc.*, **95**, 3080 (1973).
[2] B. S. Kupin and A. A. Petrov, *Zh. Obshch. Khim.*, **31**, 2958 (1961).

8.22. *TRANS*-2-METHYLCYCLOPENTYLETHYNE

3 + BH_3 $\xrightarrow[25°]{THF}$ $)_3B$

$)_3B$ + $LiC{\equiv}CH{\cdot}(CH_2NH_2)_2$ $\xrightarrow[25°]{THF}$ $[\ )_3BC{\equiv}CH\ ]\,Li{\cdot}(CH_2NH_2)_2$

$[\ )_3BC{\equiv}CH\ ]\,Li{\cdot}(CH_2NH_2)_2$ + I_2 $\xrightarrow[78°]{THF}$

$C{\equiv}CH$ + LiI + $)_2BI$

Procedure by James A. Sinclair and M. Mark Midland[1]

Procedure

A dry 500-ml three-neck flask fitted with a septum inlet, pressure-equalizing dropping funnel, large magnetic stirring bar, mercury bubbler, and side flask attached by wide flexible tubing for introduction of a solid **(S9.5)** is assembled and flushed with nitrogen. All operations are carried out under nitrogen until after the final distillation. An identical side flask is charged with 9.2 g (0.100 mole) of lithium acetylide-ethylene diamine **(N1)** in a glove bag under nitrogen and attached to the apparatus by replacing the original flask under a stream of nitrogen with minimum exposure to the atmosphere. The apparatus is again flushed with nitrogen and maintained under a static pressure of the gas. In the 500-ml flask is placed 50 ml of tetrahydrofuran and 50 ml of the 2.00*M* solution of the borane-tetrahydrofuran complex **(P2.3, P2.4:N3)**. The flask is immersed in a water bath at 25°, and 24.6 g of 1-methylcyclopentene (0.300 mole) **(P4.4:N2)** is added. The reaction mixture is stirred overnight at 25° **(N2)**. The lithium acetylide reagent in the side flask is then shaken into the solution, and the reaction mixture is stirred for 2 hours. In the dropping funnel is placed a solution of 25.4 g (0.100 mole) of iodine in 120 ml of tetrahydrofuran. The solution of the "ate" complex is cooled to −78°, stirred very vigorously, and the iodine solution is added dropwise. After 1.5 hours at −78°, the reaction mixture is brought to room temperature, and 50 ml of 40% potassium hydroxide is added. The reaction mixture is stirred for 15 minutes, and the aqueous layer is removed via syringe. Analysis of the organic phase reveals the presence of a 90% yield of *trans*-2-methylcyclopentylethyne.

The organic layer is dried by the addition of anhydrous potassium carbonate under a stream of nitrogen. The supernatant liquid is then transferred using a double-ended needle to another 500-ml flask which is connected to a simple distillation apparatus. The remaining solid is washed once with 10 ml of pentane and combined with the supernatant liquid. A 500-ml receiver flask is cooled to −78°. The volatiles are distilled under aspirator vacuum. The pot is then heated to 50° for 0.5 hours to transfer residual product **(N3)**. The assembly is returned to atmospheric pressure with nitrogen. The receiver flask is removed and connected to an efficient distillation apparatus, such as a Widmer column **(S9.7)**. The bulk of the tetrahydrofuran is distilled and then the remaining pot residue is transferred to a smaller (50-ml) flask for final distillation. Careful fractionation yields 7.15 g (66%) of *trans*-2-methylcyclopentylethyne, bp 116-117°, $n^{20}D$ 1.4425. VPC examination showed the product to be greater than 95% pure **(N4)**.

Notes

1. The lithium acetylide-ethylene diamine reacts slowly with atmospheric oxygen and moisture. It is best handled under nitrogen in a glove bag **(S9.5)**.
2. The hydroboration of this trisubstituted olefin goes rapidly to the R_2BH stage and relatively slowly beyond. In the present case completion is facilitated by the higher than usual temperature for the hydroboration, 25°.
3. This flash distillation is necessary for the separation of the product from the less volatile borinic acid. Prolonged or vigorous heating of the residue should be avoided, since this leads to decomposition to olefinic products which are difficult to separate from the acetylene.
4. Even though tetrahydrofuran and the product boil approximately 50° apart, careful fractionation is required to obtain a pure product.

[1] M. M. Midland, J. A. Sinclair, and H. C. Brown, *J. Org. Chem.*, **39**, 731 (1974).

8.23. 2-BROMO-4-METHYLHEXANAL

$$sec\text{-}Bu_3B + CH_2{=}C(Br)CHO \xrightarrow[25^\circ]{THF} CH_3CH_2CH(CH_3)CH_2CH(Br)CH{=}CH\text{-}OB(sec\text{-}Bu)_2$$

$$CH_3CH_2CH(CH_3)CH_2CH(Br)CH{=}CH\text{-}OB(sec\text{-}Bu)_2 \xrightarrow[25^\circ]{THF\text{-}H_2O}$$

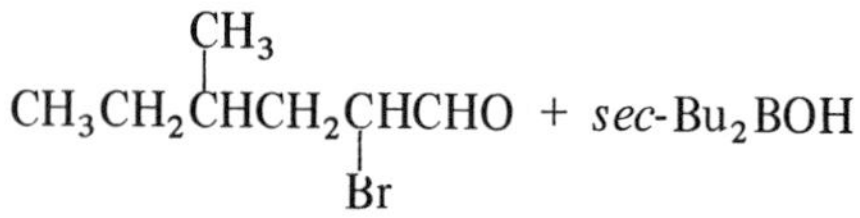

$$CH_3CH_2\underset{}{\overset{CH_3}{\overset{|}{C}}}HCH_2\underset{\underset{Br}{|}}{C}HCHO + sec\text{-}Bu_2BOH$$

Procedure by George W. Kabalka[1]

Procedure

A dry 200-ml flask fitted with a septum inlet, a condenser, magnetic stirring bar, and a mercury bubbler is assembled, flushed with nitrogen, and maintained under a static pressure of the gas until following the hydrolysis step (**S9.1**). In the flask is placed 50 ml of a 2.00*M* solution of borane-tetrahydrofuran (**P2.3, P2.4:N3**). The solution is cooled to −10° and 16.8 g (0.300 mole) of 2-butene is added as described in **P2.7**. The reaction mixture is stirred for 1 hour at 0° and then brought to room temperature. Water, 1.8 ml (0.10 mole), is added (**N1**). Then 13.5 g (0.100 mole) of 2-bromoacrolein (**N2**) is added. An exothermic reaction occurs, and the temperature rises spontaneously to approximately 40°. The reaction mixture is cooled to room temperature. Examination of an aliquot of the solution by NMR reveals a yield of 90%. Distillation provides 15.4 g (80%) of 2-bromo-4-methylhexanal, bp 60° (4.8 mm) (**N3, N4**).

Notes

1. When commercial borane-tetrahydrofuran is used, 2.0 ml of 3.0*M* hydrochloric acid is used in place of the water to destroy the sodium borohydride stabilizer.
2. In a 1-ℓ flask is placed 56 g (1.0 mole) of acrolein dissolved in 500 ml of water. The flask is cooled to 0° by an ice-water bath, and 160 g (1 mole) of bromine is added slowly over a period of 3 hours, maintaining the temperature at 0-5°. After the addition has been completed, the reaction product is subjected to steam distillation. The 2-bromoacrolein layer (lower) is collected and distilled: 55 g [41% yield of 2-bromoacrolein, bp 45° (25 mm)].[2]
3. The α-bromoaldehydes are exceedingly reactive and cannot be stored as such for any length of time. They are readily converted into diethylacetals[1] and may even be isolated in that form from the reaction mixture in higher yields (~85%).[1]
4. An alternative approach to α-halocarbonyl derivatives involves treatment of the enol borinates with *N*-halosuccinimide.[3]

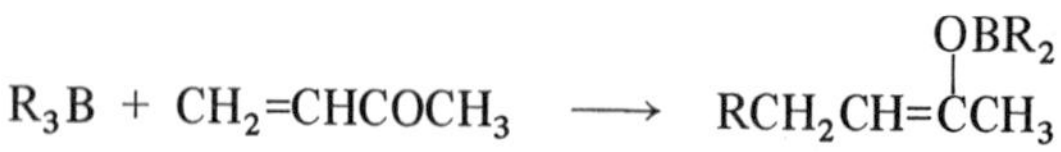

$$R_3B + CH_2{=}CHCOCH_3 \longrightarrow RCH_2CH{=}\overset{\overset{OBR_2}{|}}{C}CH_3$$

$$RCH_2CH{=}\overset{OBR_2}{\overset{|}{C}}CH_3 + \text{NBS} \longrightarrow RCH_2\underset{Br}{\underset{|}{C}}HCOCH_3 + \text{N-}BR_2\text{-succinimide}$$

[1]H. C. Brown, G. W. Kabalka, M. W. Rathke, and M. M. Rogić, *J. Amer. Chem. Soc.*, **90**, 4165 (1968).
[2]A. Berlande, *Bull. Soc. Chim. France*, 37[4], 1385 (1925).
[3]J. Hooz and J. N. Bridson, *Can. J. Chem.*, **50**, 2387 (1972).

8.24. 5,5,6-TRIMETHYL-2-HEPTANONE

$$\text{3,5-dimethylborinane (B–H)} + (H_3C)_2C{=}C(CH_3)_2 \xrightarrow[25^\circ]{THF} \text{B-thexyl-3,5-dimethylborinane}$$

$$\text{B-thexyl-3,5-dimethylborinane} + CH_2{=}CHCOCH_3 + H_2O \xrightarrow[25^\circ]{THF} H\text{–}C(CH_3)_2\text{–}C(CH_3)_2\text{–}CH_2CH_2COCH_3 + \text{B(OH)-3,5-dimethylborinane}$$

Procedure by Ei-ichi Negishi[1]

Procedure

A dry 200-ml three-neck flask equipped with a septum inlet, condenser, magnetic stirring bar, and mercury bubbler is assembled, flushed with nitrogen, and maintained under a static pressure of the gas until following the oxidation **(S9.1)**. In the flask 0.04 mole of 3,5-dimethylborinane is prepared as a 1.00*M* solution **(P2.12)**. To this solution is added 3.4 g (0.040 mole) of 2,3-dimethyl-2-butene **(S9.11)**, and the reaction mixture is allowed to stir for 16 hours or longer **(N1)**. To the reaction mixture is added 1.5 ml (0.080 mole) of water. After hydrogen evolution ceases, 4.2 g (0.050 mole) of methyl vinyl ketone. The reaction mixture is stirred overnight at 25°. VPC examination of the reaction mixture reveals the presence of an 88% yield of 5,5,6-trimethyl-2-heptanone **(N2)**. The intermediate is oxidized to facilitate isolation of the product. To the flask is added 13.3 ml of 3*M* sodium hydroxide, followed by the slow dropwise addition of 8.4 ml of 30% hydrogen peroxide (exothermic! **P2.1:N6**). Sodium chloride is added to saturate the aqueous phase. The tetrahydrofuran layer is separated, and the aqueous layer extracted with pentane (2 X 20 ml). After drying over anhydrous magnesium sulfate, the combined organic layer is distilled to yield 4.6 g (75% yield) of 5,5,6-trimethyl-2-heptanone: bp 95-97° (20 mm), $n^{20}D$ 1.4356.

Notes

1. Most olefins undergo hydroboration with 3,5-dimethylborinane under these conditions within 1-2 hours. Cyclohexene requires approximately 8 hours and 2,3-dimethyl-2-butene approximately 16 hours.
2. Generally the conjugate addition of methyl vinyl ketone and organoboranes proceeds spontaneously. In some cases the introduction of a small amount of air with a hypodermic syringe is required to initiate the reaction, even with methyl vinyl ketone.[3]
3. B-*t*-butyl-3,5-dimethylborinane (**P4.8**) transfers the *t*-butyl group in such conjugate additions.[1]

$$CH_2{=}CHCOCH_3 \longrightarrow H_3C{-}C(CH_3)_2{-}CH_2CH_2COCH_3$$

CH3 / CH3 groups on central C; cyclohexenone $\xrightarrow{\text{cat. } O_2}$ 3-(C(CH3)3)cyclohexanone

[1] H. C. Brown and E. Negishi, *J. Amer. Chem. Soc.*, **93**, 3777 (1971).
[2] E. Negishi and H. C. Brown,. *J. Amer. Chem. Soc.*, **95**, 6757 (1973).
[3] H. C. Brown and G. W. Kabalka, *J. Amer. Chem. Soc.*, **92**, 714 (1970).

GARY W. KRAMER, ALAN B. LEVY AND M. MARK MIDLAND 9

LABORATORY OPERATIONS WITH AIR-SENSITIVE SUBSTANCES: SURVEY

The sensitivity of most organoboron compounds toward air, and in some cases water, has no doubt led some workers to hesitate making use of their remarkable chemistry. Recalling the prehydroboration era, some may still believe that very specialized apparatus is necessary to carry out reactions involving diborane and organoboranes. This is far from true today. In his recent book, *The Manipulation of Air-Sensitive Compounds,* D. F. Shriver has done an outstanding job of presenting many techniques for handling air-sensitive materials.[1] It is our aim to complement his book by describing the techniques we have found most useful. Therefore, we do not describe many of the more specialized types of inert-atmosphere techniques, such as Schlenk-type apparatus, vacuum lines, or inert-atmosphere boxes, but concentrate our efforts on presenting simple, but adequate, bench-top methods which have proved to be highly satisfactory in the organoborane area. We describe techniques which utilize common laboratory equipment and which can be used to carry out, on a preparative scale, all of the reaction sequences described earlier. We also describe some of the more sophisticated apparatus that is used routinely for small-scale developmental reactions in our laboratories. We limit our discussion to techniques which are unique to working with air-sensitive materials and make little effort to describe practices which should be common knowledge to most organic chemists. These techniques are adequately described elsewhere.[2,3]

[1]D. F. Shriver, *The Manipulation of Air-Sensitive Compounds,* McGraw-Hill, New York, 1969.
[2]K. B. Wiberg, *Laboratory Technique in Organic Chemistry,* McGraw-Hill, New York, 1960.
[3]A. Weissberger (ed.), *Technique of Organic Chemistry,* Wiley-Interscience, New York, 1963.

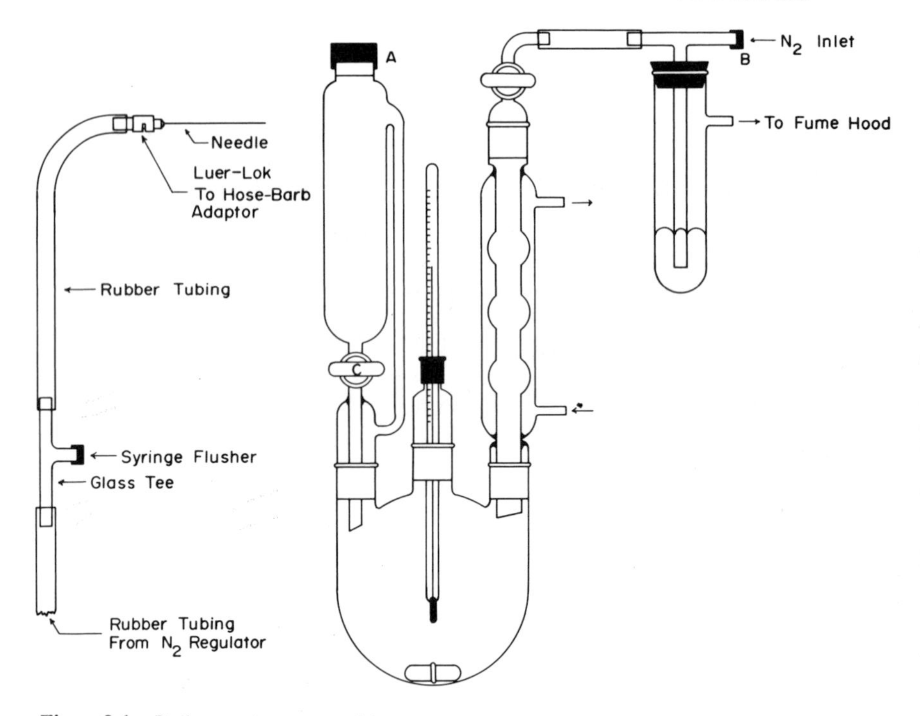

Figure 9.1. Basic reaction setup with common glassware.

9.1. APPARATUS FOR REACTIONS INVOLVING AIR-SENSITIVE MATERIALS

General Apparatus

Reactions involving oxygen- and/or water-sensitive materials may be carried out in common ground-glass apparatus (**F9.1**). The only additional equipment required is a source of inert gas (**F9.1**, **F9.9**), a septum-capped inlet (**F9.2-F9.6**), a mercury or oil bubbler (**F9.7**), and hypodermic syringes and needles (**S9.2**).

A typical apparatus for carrying out a hydroboration-oxidation sequence (**P2.4**) is shown in **F9.1**. The glassware may either be predried in an oven (4 hours at 150° or 12 hours at 125°), assembled hot, and allowed to cool under a stream of nitrogen; or it may be assembled and flame dried under a stream of nitrogen. A simple nitrogen source can be constructed from medium-wall rubber tubing connected at one end to the gas cylinder through a pressure regulator (set to 3-5 psi) and on the other end to a 6-in. 20-gauge hypodermic needle. A septum-capped tee can be spliced into the line to permit flushing syringes with nitrogen.

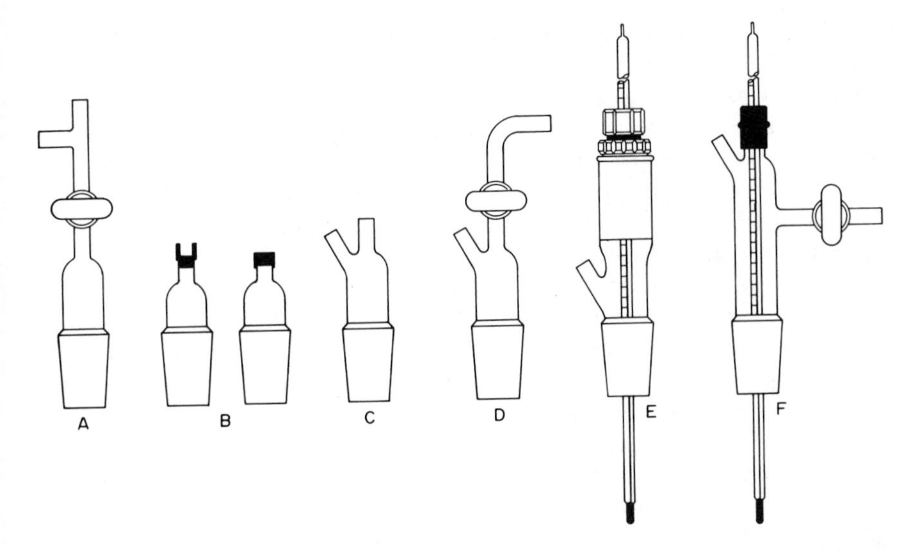

Figure 9.2. Septum-inlet adapters.

The glassware is assembled, and the nitrogen needle is inserted through septum A **(F9.1)**. If the glassware is not predried, the entire apparatus is flamed out with a Bunsen burner. After the system has cooled and has been thoroughly flushed with nitrogen, the needle is removed from septum A and inserted through septum B on the bubbler. This maintains a static, positive pressure of nitrogen on the apparatus and prevents the bubbler liquid from being sucked back should a pressure reversal occur during the reaction. At this point all of the ground joints should be securely fastened together with rubber bands, spring clips,[4a] or other clamps[4b] to prevent leakage.

In a typical hydroboration **(P2.4)**, the reaction flask is charged with liquid olefin from a syringe through septum A and the addition funnel. The solvent is added in the same manner. Stopcock C is closed, and the proper amount of BH_3:THF is added to the addition funnel through septum A via syringe. The hydroboration reaction is carried out by adding the BH_3:THF dropwise to the olefin solution. When this reaction is complete, first water and then aqueous base are added by syringe through septum A and the dropping funnel. Stopcock C is closed, and the hydrogen peroxide solution is added by syringe through septum A to the dropping funnel. The oxidation step is carried out by adding the hydrogen peroxide dropwise to the reaction mixture (exothermic reaction!).

[4a]Available from Ace Glass, Inc., or A. H. Thomas and Co.
[4b]Available from Kontes and other laboratory supply houses.

Following the oxidation, the apparatus may be disassembled and the product isolated by conventional means.

Most reactions involving organometallics, including all of the procedures described here, can be carried out in common ground-glass apparatus using suitably designed adapters with septum inlets (**F9.2**).[5] They have an advantage over the large septum in **F9.1** in being more conveniently located for the introduction and removal of reagents. The smaller septum also is safer to use in operations requiring evacuation.

F9.3 shows an apparatus utilizing an adapter (**F9.2**) equivalent to the apparatus previously discussed (**F9.1**). This apparatus offers several advantages. The additional septum inlet D allows the apparatus to be more easily flushed with nitrogen. After flushing through septum A, the needle can be inserted through septum D to the bottom of the flask. Nitrogen, which is slightly less dense than air, will displace the air more efficiently if the inlet needle reaches the lowest point of the apparatus. Reagents may be added to the reaction mixture through septum inlet D without contaminating the addition funnel. Using adapters similar to those described (**F9.2**) will be the method of choice for the occasional user.

Special Glassware

Routine work with air-sensitive materials easily justifies the use of modified glassware. Such specialized glassware can be constructed from standard ground-glass equipment by blowing on a 9-mm medium-wall sidearm for a septum inlet (**F9.4-6**). **F9.4** shows a modified apparatus (**P2.4**) equivalent to that depicted in **F9.1**. Although it is used in an analogous manner to that previously described, this apparatus offers certain advantages. The additional septum D on the flask facilitates removal of aliquots for VPC analysis or other purposes. In addition, it allows all of the operations possible with the equipment described in **F9.3**.

In cases where the condenser and dropping funnel are not required (**P2.9**), a flask equipped with a septum-capped inlet and a stopcock-controlled connecting tube will usually suffice (**F9.5**). When it is desirable to store a sensitive product directly after the reaction without transferring it, the apparatus shown in **F9.6** is suitable (**P2.11**). After the completion of the reaction, a nitrogen needle is inserted through the stopcock-controlled septum inlet A. With nitrogen flowing, the condenser and the addition funnel are removed, and the stopcock-controlled connecting tube is attached directly to the flask. Aliquots of the reaction mixture can be easily removed from the flask by syringe through the septum-capped inlet. The stopcock keeps solvent and solvent vapors away from the septum, thereby minimizing deterioration and swelling of the septum. Further-

[5]Many of these adapters are available from Aldrich Chemical Company.

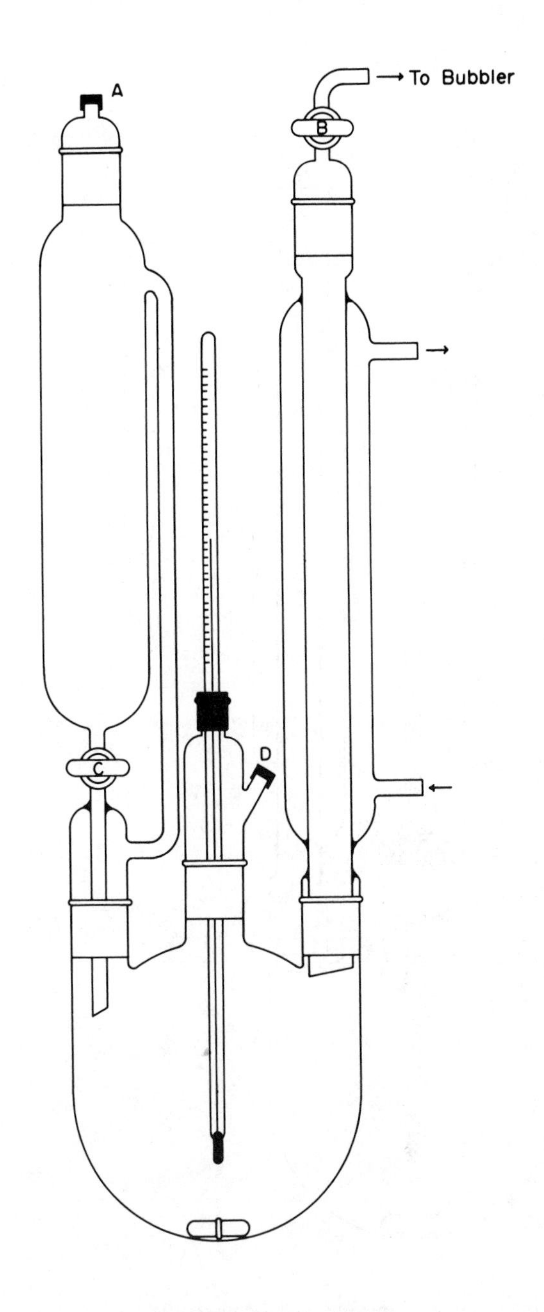

Figure 9.3. Basic reaction setup using septum-inlet adapters.

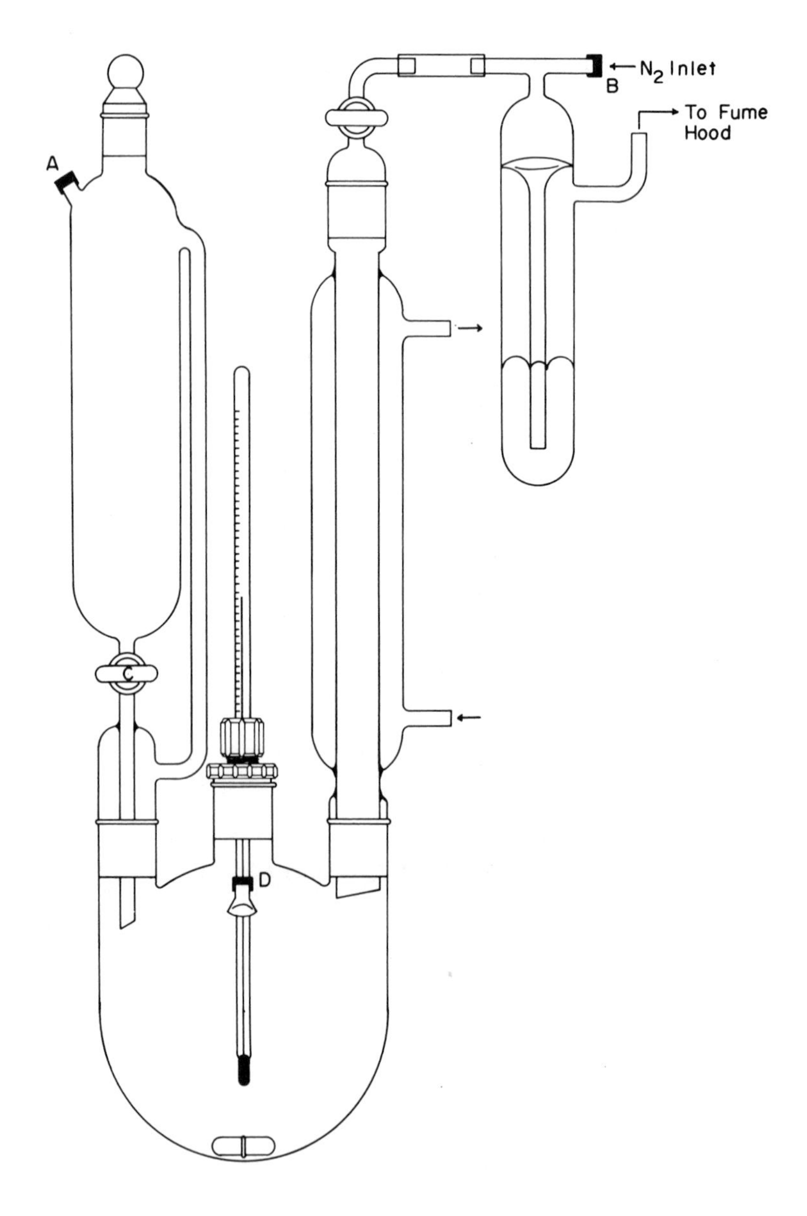

Figure 9.4. Basic reaction setup with modified glassware

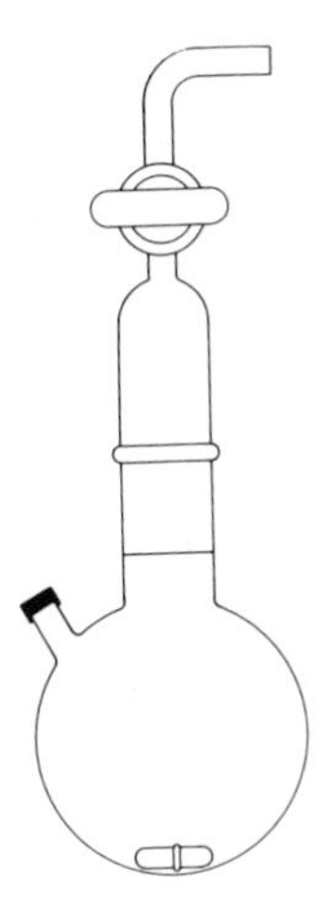

Figure 9.5. Simple reaction flask.

more, on standing for long periods, rubber septa may deteriorate due to chemicals in the atmosphere. A stopcock prevents oxygen or moisture from diffusing through a leaking septum and makes it possible to replace a defective rubber septum with minimum exposure of the contents to the atmosphere.

Stirrers

All the equipment presented is designed to be stirred magnetically using Teflon-coated stirring bars. Generally, magnetic stirrers are adequate for the procedures described. Occasionally, it is desirable to shift to an overhead mechanical stirrer to handle highly viscous products, suspended solids, very large-scale preparations, or high-speed stirring. If a mechanical stirrer is required, a precision-ground stirrer,[6] sealed at the top with mercury or oil, is usually sufficient. If large pressure differentials are anticipated, stirrers with stuffing box or O-ring seals are preferred.

Bubblers and Waste Line

Mercury or oil-filled bubblers **(F9.7)** are used in inert-atmosphere reaction setups to provide a gas seal between the reaction vessel and the atmosphere. Often a pressure reversal in the reaction vessel can cause the liquid in the bubbler to be sucked back. A tee-tube bubbler **(F9.7A)** can be used to prevent this problem by allowing a slow stream of nitrogen to flow continuously through the bubbler.

[6]Available from Ace Glass, Inc. as a Trubore stirrer.

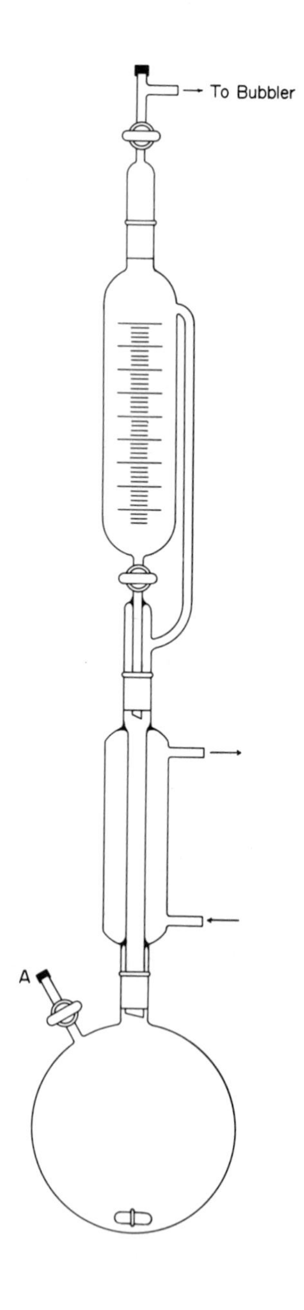

Figure 9.6. Reaction setup with stopcock-controlled inlet. The condenser and dropping funnel may be replaced with adapter **F9.2A** for storage of air-sensitive products.

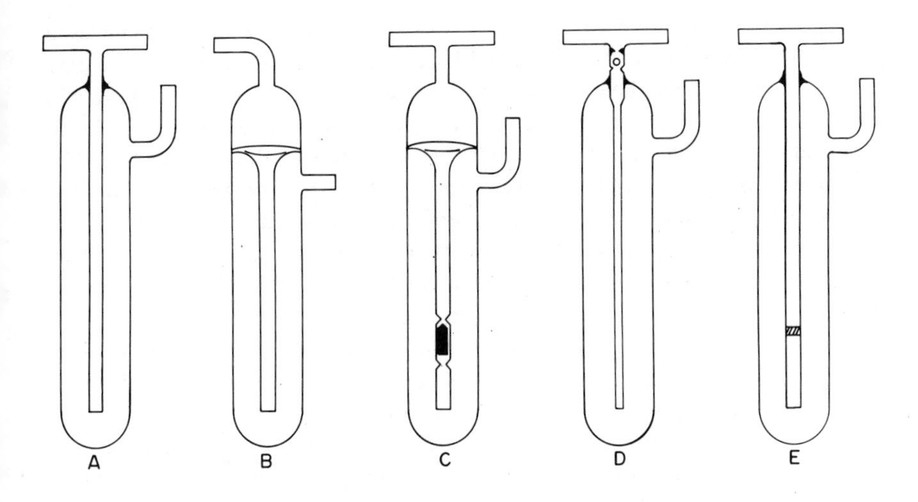

Figure 9.7. Typical bubbler design: (A) tee-tube bubbler; (B) Ace Glass Cat. No. 8761-10; (C) tee-tube bubbler with check valve; (D) tee-tube bubbler with ball check valve; (E) tee-tube bubbler with medium-porosity fritted disk.

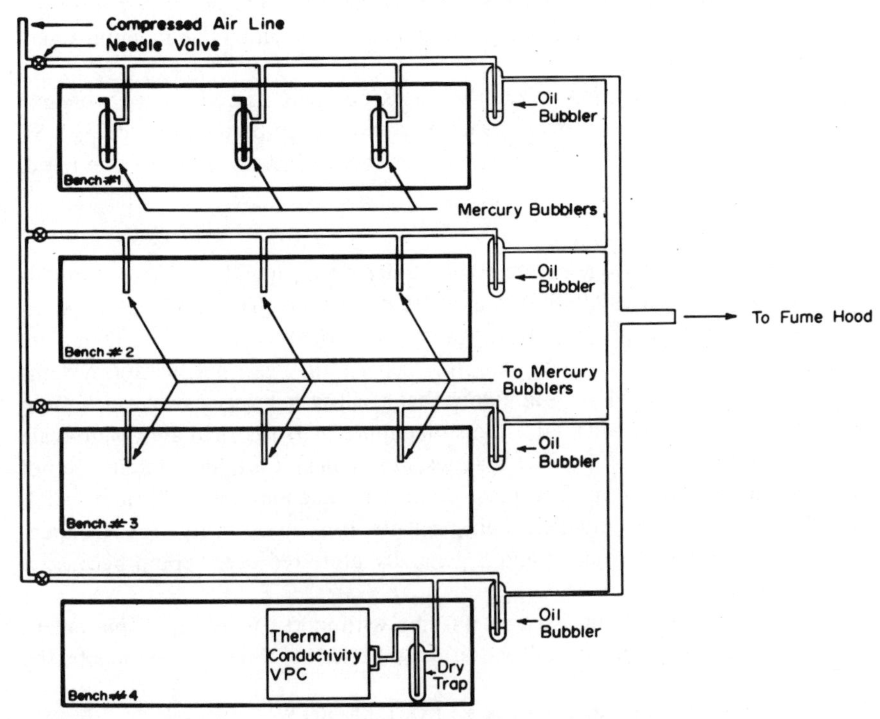

Figure 9.8. Waste-gas disposal system for multibench laboratory.

However, over a long period of time a considerable amount of nitrogen is used. Commercial bubblers with an enlarged head space (**F9.7B**) minimize the danger of the bubbler liquid being sucked back into the reaction vessel. However if a pressure reversal occurs, air will be admitted to the reaction setup. A better solution to this problem is to use a bubbler with a built-in check valve (**F9.7C-E**). The valve can be a ground-glass check valve (**F9.7C,D**), or when using mercury as the bubbler liquid, a medium-porosity fritted disk will serve as a one-way valve (**F9.7E**). A disadvantage of a fritted disk is that the frit may become plugged. These types of bubblers eliminate both liquid suck back and the intrusion of air while still providing for the release of excessive pressure.

The effluent gas from the bubbler should be piped to a fume hood, or when using mercury, through a mercury scrubber and then to the fume hood (**F9.8**). It is convenient to gang the outlets from all the bubblers on one bench together and pass only one line to the fume hood. To insure that any vapors present reach the hood, this line can be continuously swept with air (5-10 ml/minute). In large laboratories, the lines from each bench can be connected to a central pipe leading to the hood. In this way the problem of having many small tubes leading into the hood can be avoided. Each bench line may be isolated from the central pipe with an oil bubbler to prevent the effluent gas from one bench reaching another. Effluent from gas chromatographs and vacuum pumps may also be connected to the waste line.

A waste line greatly minimizes the odors normally associated with organic laboratories. In addition, many reactions, such as carbonylation (**P8.3, 8.5**), can be safely carried out on the laboratory bench instead of in a fume hood.

Inert Gases

Most organometallic reagents are stable under nitrogen and for these materials prepurified nitrogen (99.998%) is a suitable inert gas. This grade of nitrogen can be used without further purification for all of the procedures described here. For more sensitive compounds requiring higher purity gas, purification systems are available.[7] It should be pointed out that small-scale reactions are much more susceptible to the detrimental effects of impure inert gas than are macro-scale preparations (one drop of water is almost 3 mmoles). Certain materials are not compatible with nitrogen. The most commonly encountered is lithium metal, which is used to make organolithium reagents. In such cases argon is an expensive, but necessary alternative and is generally preferred over helium because it is heavier than air.

When the reaction setup is large, flushing with inert gas by the displacement method is time-consuming and wasteful. A better solution is to evacuate the

[7a]For a discussion of purification of gases, see Ref. 1, Chapter 9.
[7b]A commercial gas purification system is available from Ace Glass, Inc.

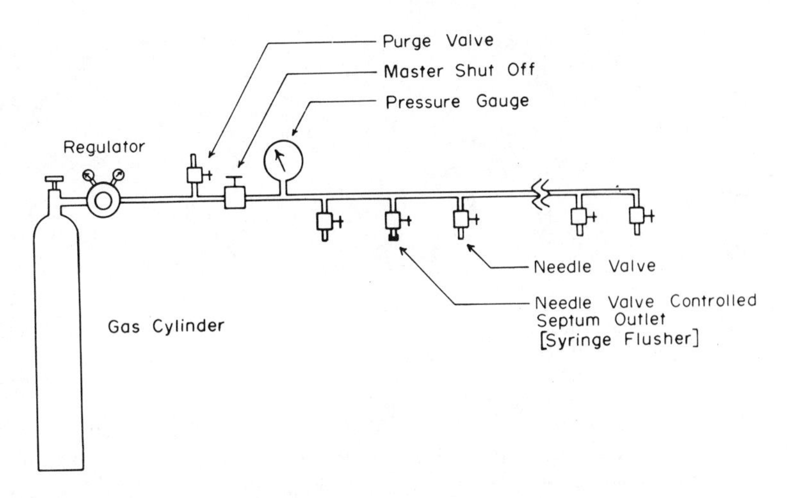

Figure 9.9. Metal nitrogen manifold. A syringe flusher may be provided by drilling out a hose-barb to allow capping with a 6-mm septum.

apparatus with an oil pump and then refill the system with nitrogen. The purge cycle is repeated to ensure the complete removal of oxygen. This method is especially advantageous in conjunction with drying by flaming. The evacuated system is flamed out and then refilled with nitrogen. The drying/purge cycle is generally repeated. If the system is connected to a bubbler with a built-in check valve (**F9.7C-E**), the system may be evacuated and then refilled without the danger of overpressurizing the system during the refill.

Frequent work with sensitive materials justifies a more sophisticaled nitrogen system than that described earlier (**F9.1**). Since rubber and plastic tubing deteriorate in most laboratory environments, nitrogen manifolds constructed of these materials require periodic replacement to avoid leakage. An all-metal nitrogen system constructed from 3/8- or ¼-in. galvanized pipe is functionally and economically more desirable for a permanent installation (**F9.9**).

When building such a system, the pipe is cut to the desired lengths and threaded. The pipes and fittings are washed with solvent to remove any oil or dirt inside. After drying, the line is assembled using Teflon tape to seal the threads. Needle valves are used on the outlets to control the gas flow. Lengths of rubber tubing are used to connect the needle valves to hypodermic needles attached via a Luer-Lok to hose-barb adapter. When not in use, the needles are pushed into small rubber stoppers to keep air out of the rubber tubing. At least one needle-valve-controlled septum outlet should be provided for use in flushing syringes. The pressurized system should be leak tested using a soap solution on the joints. Small leaks may be detected by pressurizing the system to about 40 psi and closing off the main cylinder valve. A good system will

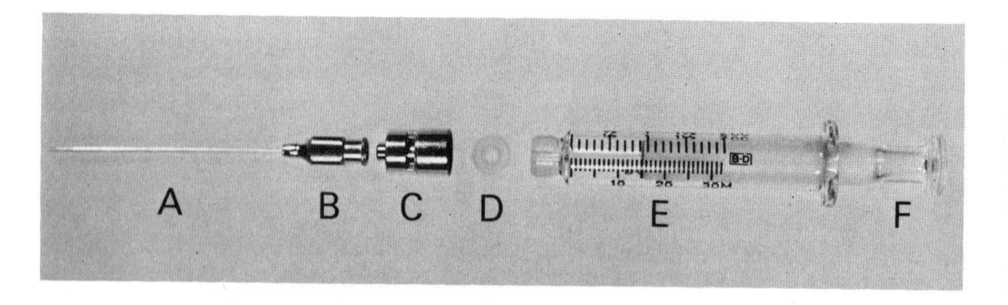

Figure 9.10. Exploded view of a hypodermic syringe: (A) cannula; (B) needle hub; (C) Luer-Lok tip; (D) sealing washer; (E) barrel; (F) plunger.

maintain the pressure indefinitely. The nitrogen line is purged of air by evacuation with an oil vacuum pump followed by filling with nitrogen. This purging cycle should be repeated several times. Residual moisture in the line may be eliminated by heating the pipe overnight to about 80°C with heating tape while evacuating with an oil pump. If a Bourdon pressure gauge is used on the line, it should be removed before the line is evacuated to prevent damage. The gauge may be installed, after evacuation, with nitrogen flowing through the system.

9.2. LIQUID TRANSFER TECHNIQUES

Hypodermic syringes offer a convenient method for transferring air-sensitive liquids. Although the operation of a hypodermic syringe appears to be self-evident, it is our experience that the proper utilization can be greatly facilitated by certain background information. It is the purpose of this section to impart this background.

Syringes

Standard hypodermic syringes (**F9.10**) are made in sizes from ¼ ml to 200 ml by Becton-Dickinson and Co. They are generally classified by their medical usage and are available from most chemical or hospital supply houses. The barrel is available with three tip styles: metal Luer tips, glass Luer tips, and needle-locking tips (Luer-Lok) (**F9.11**). A Luer-Lok tip offers the advantage of securely holding the needle to the syringe, and preventing it from loosening during use. It should be noted that the metal tips are chrome-plated brass; consequently, they are subject to corrosion by certain materials.

Small syringes, ¼ to 1 ml, are called Tuberculin syringes. These are an example of a precision syringe. Such syringes have individually mated plungers and barrels identified by matching code numbers on both parts. It is important to realize that these parts are *not* interchangeable from syringe to syringe.

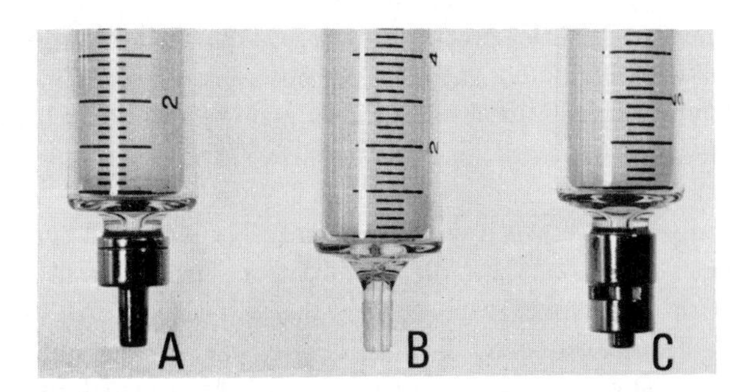

Figure 9.11. Syringe tip styles: (A) metal Luer; (B) glass Luer; (C) Luer-Lok.

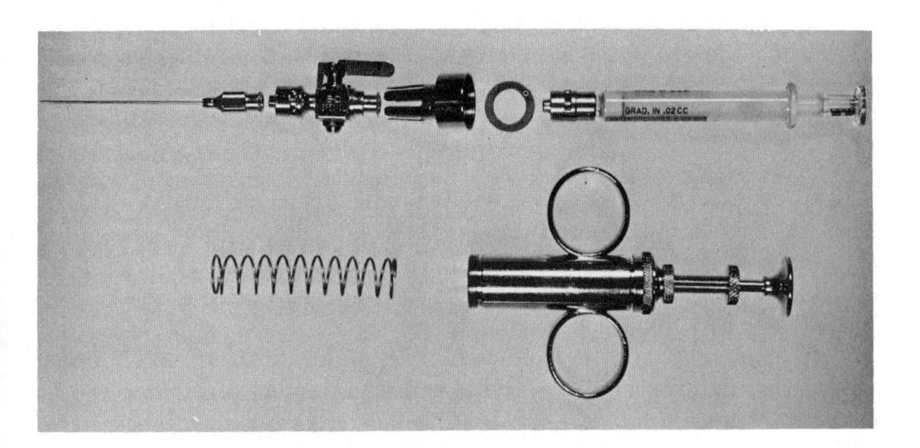

Figure 9.12. Exploded view of a Cornwall syringe.

Interchanging these pieces can cause leaking, sticking plungers, or miscalibration.

Larger syringes are generally of the Multifit variety. Here plungers and barrels of a given size syringe are interchangeable. These syringes are identified by an encircled *M* or the word, Multifit, on the plunger or barrel.

Plastic, disposable syringes offer an inexpensive method for transferring aqueous solutions, such as 30% hydrogen peroxide or 3*M* sodium hydroxide. Since they cannot be oven-dried, and since organic solvents frequently cause their rubber plungers to swell, they are not generally suited for use with organometallic reagents.

Microliter syringes with capacities between 1 μl and 500 μl are available from the Hamilton Co. and other suppliers. Although they are primarily used for chromatographic injections, they are useful for transferring small amounts of air-sensitive compounds. Like precision syringes, the metal plungers and glass barrels are individually mated and are *not* interchangeable. They should be marked to avoid mixups. Microliter syringes may be ordered with extra long needles. Needles 7½-in. long are particularly useful for reaching to the bottom of standard 7-in. NMR tubes.

There are many syringes with special designs. The Cornwall syringe and the gas-tight syringe are two of the more useful types. The Cornwall syringe has a metal casing with finger loops which covers the plunger and part of the syringe barrel (**F9.12**). This cover serves as a holder for the syringe and prevents the plunger from popping out. When used with a syringe stopcock on the tip, the syringe can be operated with one hand.

Gas-tight syringes are available in sizes from 50 μl to 1.5 ℓ. The smaller sizes, up to 50 ml, have glass barrels and metal plungers with Teflon tips. Although they are designed for the transfer of gases, they are useful for manipulating volatile liquids, such as pentane or ether, which tend to leak in ordinary syringes.

Needles

Syringe needles are available in a myriad of designs, sizes, and point styles. The most useful type is the Local Anesthesia needle. These generally have 20- or 22-gauge cannulas in lengths up to 6 in. The hub is chrome-plated brass, while the cannula is stainless steel. For details on needles in other sizes or made of other materials, such as Kel-F or Teflon, the reader is referred to the Becton-Dickinson or Hamilton catalog.

Three point styles are especially useful: a standard bevel, 17° or 12° taper, for normal septum penetration; Huber points for septum penetration with a minimum of coring (making a nonresealable hole in the septum); flat-cut bevel (90°) needles for syringe pipetting or getting the last few drops of liquid out of a flask (**F9.13**). In general, long needles seem to work best because they can be bent so that the syringe can be inverted after filling to remove entrapped gas bubbles while the tip of the needle is still in the flask. Short, small-bore (20-26 gauge) needles are well suited for gas venting purposes.

Often the user may wish to modify the length or tip style of a needle. The cannula may be cut by scoring the metal with a triangular file and then snapping the metal (much like cutting glass tubing). It may also be cut with diagonal cutters. When using the latter method, cutting the tubing while a cleaning wire is inserted in the needle will prevent the collapse of the adjacent cannula. After cutting, the point is first filed roughly to the style desired. This is easily done if the file is held stationary in a bench vise and the needle, held in a pin vise, is moved. The proud metal which forms inside the cannula can be removed with the aid of a cleaning wire. After the rough shaping, the point is sharpened on an oil stone (much like sharpening a knife). If the sharpening is not done well, the needle will core septa badly. A properly beveled and sharpened needle will

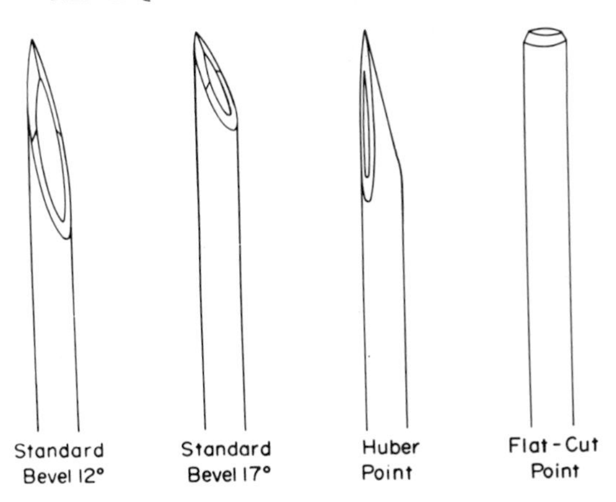

Figure 9.13. Cannula point styles.

leave only a small slit in a septum. When using a flat-cut needle, septum coring can be minimized by prepuncturing the septum with a standard-beveled needle. The flat-cut needle is then inserted through the existing puncture mark. In general, it is best to use existing puncture marks when reinserting any needle, since this prolongs the life of the septum.

Syringe Fittings and Accessories

The user should be aware of the variety of syringe related hardware that is available in metal fittings from Becton-Dickinson or in Kel-F from Hamilton. These include stopcocks, needle plugs, syringe caps, hose barb to Luer adapters, Luer to Luer adapters, and Luer to standard threaded connectors. Also available are connectors for attaching syringe fittings to Lecture Bottles. These will be described in **S9.9**.

Periodically it is convenient to fabricate glassware with syringe fittings. This is easily accomplished since glass Luer fittings are available from laboratory glass supply houses. Three sizes are available: the small size is the normal Luer taper, the middle size is the large Luer taper found on some large syringes, the largest size is the taper to which the metal Luer-Lok is fastened. Glass apparatus should not be constructed from broken syringes because they are not normally made of borosilicate glass.

Metal Luer-Lok fittings are held onto the glass by a contraction fit. No adhesive agent is used. To remove a Luer-Lok fitting, one first fastens a large-bore needle into the fitting to use as a handle. The metal tip is then quickly heated in a flame. When the metal expands, the tip may be easily removed. The metal tip must not be overheated for this can cause distortion, damage the chrome plating, or destruction of the rubber sealing washer **(F9.10)**. A metal Luer-Lok

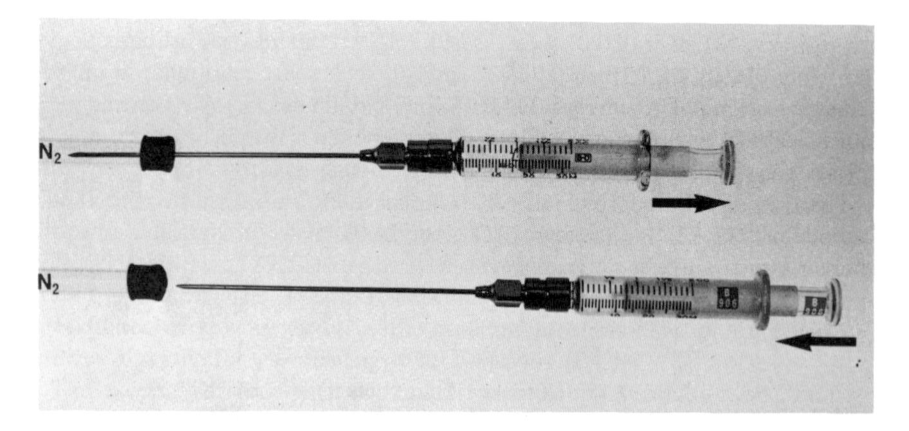

Figure 9.14. Purging a syringe.

fitting is installed by reversing the above procedure. The rubber sealing washer should be lined up on the glass tip and the heated metal tip pushed securely onto the glass. Moderate pressure is maintained on the assembly until the metal cools.

Procedures for Using Syringes

Prior to use with air- and water-sensitive materials, the syringe plunger, barrel, and needle should be disassembled and dried in an oven. The dry syringe is removed from the oven and assembled hot. The hot needle may be attached to syringes having metal tips; however, this is not generally done with glass-tipped syringes since the needle hub contracts on cooling and freezes to the glass tip. Reheating the needle hub is often necessary to remove the needle without breaking the syringe.

When the syringe is cool, a *small* amount of stopcock grease or a *drop* of silicone oil is placed on the Luer tip as a sealant, and the needle is attached to the syringe. The needle hub may be tightened with pliers. Overtightening should be severely discouraged as this permanently distorts the fitting and will cause leakage in the future.

After assembly, the needle is flushed several times with nitrogen, as shown in **F9.14**. During the flushing procedure, the syringe assembly should be leak tested by filling the syringe with gas and inserting the needle tip into a rubber stopper. The plunger is depressed about halfway and released. If there are no leaks, the plunger should return to its original position.

It is sometimes desirable to lubricate the plunger and barrel with a thin film of mineral oil or Amojell (Vasoline) to prevent the plunger from freezing. When handling moisture-sensitive compounds, such as organolithium reagents, Grignards, or boron hydride solutions, the plunger sides should not be touched since residual moisture may cause the plunger to stick.

Two methods are commonly used to dispense precise amounts of liquid from a syringe. The simplest method is to rely on the graduations on the syringe barrel. This rather crude method may be improved by calibrating the syringe. This can be done by weighing a volume of liquid delivered by the syringe. A liquid with a well established density at ambient temperature should be used. The calibration should be checked at various points on the syringe. If the syringe is properly calibrated, and if the user is competent, precisions and accuracies of ±2% can be expected. A more satisfactory procedure is to weigh the syringe before and after the liquid is dispensed. The contents of the syringe may be protected from the atmosphere or from accidental leakage during the weighing procedure by inserting the tip of the needle into a small rubber stopper.

Two techniques for transferring liquids by syringe will be described: transfer of nonsensitive or moderately sensitive liquids, such as solutions of normal trialkylboranes or BH_3:THF; transfer of very sensitive compounds, such as neat triethylborane, neat trialkylaluminums, or *tert*-butyllithium in pentane. It is assumed that the user has the liquid in a septum-capped storage vessel and wishes to transfer an aliquot into a septum-capped receiver.

First the storage vessel is pressurized (2-5 psi) with nitrogen. The receiver is flushed with nitrogen and connected to a bubbler. The syringe assembly is flushed and leak tested as previously described. The syringe is withdrawn from the flushing assembly about three-quarters full of nitrogen, and the needle is wiped free of fingerprints with a tissue. The needle is given a final purge by depressing the plunger halfway and is then inserted through the septum of the storage vessel. If the needle is grasped with the fingers during insertion, it should be wiped clean before inserting it fully. Once inside the septum, the remainder of the nitrogen is dispelled. The needle is placed below the surface of the liquid, and the pressure above the liquid is allowed to fill the syringe 10-20% above the desired level. The needle is then raised above the liquid level. Bending the needle gently, the syringe is inverted so that the entrapped gas bubbles rise to the top. These bubbles are dispelled by depressing the plunger until the correct volume remains in the syringe. Most syringes have an engraved line on the plunger tip. This line, not the bottom of the plunger, should be used as the reference point. The needle cannula is then given a flip with the finger to knock off the drop at the needle tip, and the plunger is withdrawn partially, drawing in a blanket of gas and clearing the needle. With the syringe still inverted, the needle is withdrawn from the storage vessel and inserted through the septum of the receiver (or into a rubber stopper if the weighing procedure is used). Once the needle is inserted into the receiver, the contents of the inverted syringe are discharged. It is important that the gas be dispelled before the liquid. Failure to do this will result in adding an incorrect amount of liquid. Because the syringe

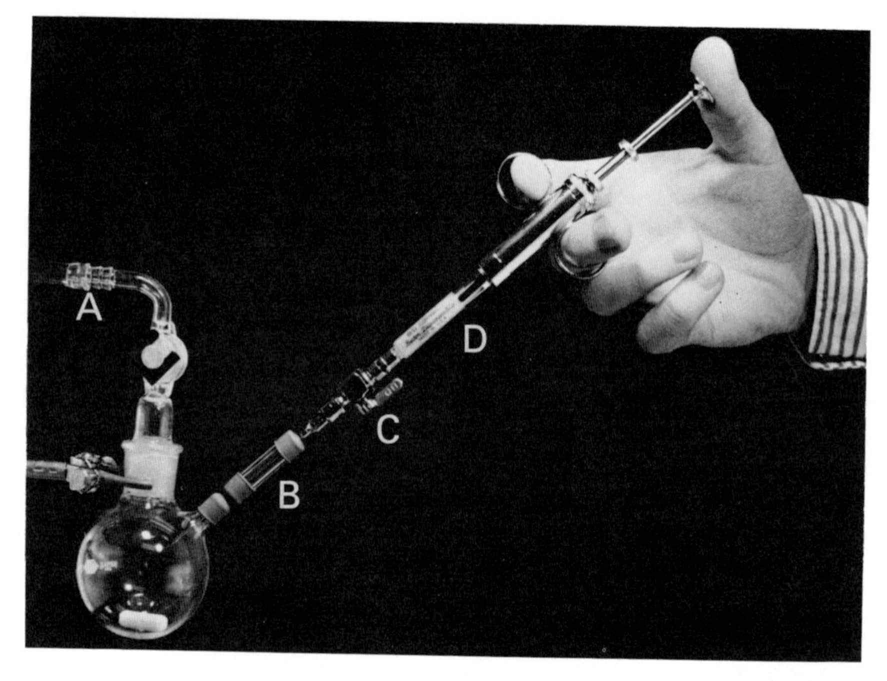

Figure 9.15. Cornwall syringe in use: (A) to bubbler; (B) 9-mm medium-walled tubing capped with 6-mm septa; (C) Luer-Lok valve (B-D No. 3152); (D) Cornwall syringe.

is a "to deliver" device, the needle must remain filled with liquid when the syringe is emptied. A small amount of inert gas is then sucked into the syringe before the needle is removed from the receiver. If the syringe is to be reused soon, the needle tip is pushed into a rubber stopper.

The procedure for transferring very sensitive compounds is quite similar to that described above. In this case a Cornwall syringe holder fitted with a syringe stopcock is preferred. Small fires or solvent evaporation at the needle tip can be minimized by using a short (½-in.) length of 9-mm medium-wall tubing fitted with two 6-mm septa **(F9.15)**. This tube is dried, fitted with the septa, and then flushed with nitrogen. The syringe needle is inserted through one of the septa. The face of the other septum is covered with stopcock grease. This septum is butted directly against the septum on the storage vessel and the needle pushed through both septa. The syringe is filled as before, and when the inert blanket of gas has been pulled in to clear the needle, the stopcock on the syringe is closed. The needle is now withdrawn so that its tip is inside the glass tube. The procedure is reversed to transfer the liquid into the receiver. This short double-capped tube is especially convenient if a needle becomes clogged while the syringe is filled with liquid. A cleaning wire can be inserted through the

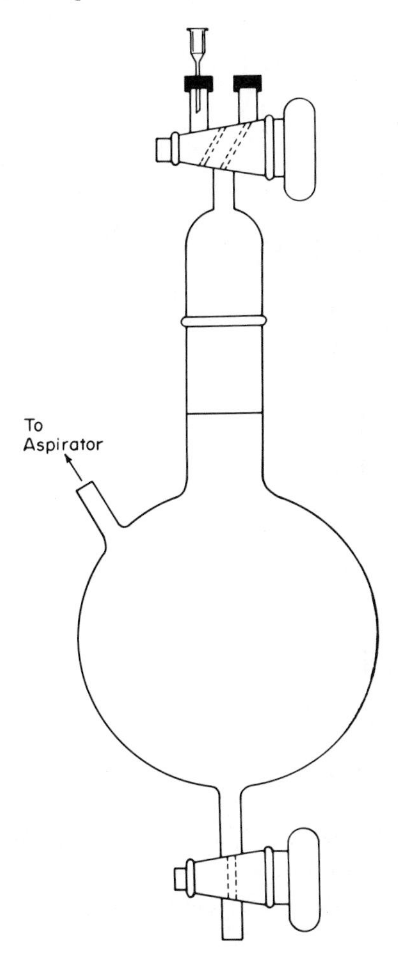

Figure 9.16. Syringe cleaner.

bottom septum to clean the needle without the danger of allowing air to come in contact with the liquid.

If a liquid must be added to a reaction mixture at known rate or over a long period of time, the syringe may be driven with a mechanical syringe pump. Several models are avilable from the Sage Instrument Co.

Syringes and needles are easily cleaned using the device shown in **F9.16**. Needles are cleaned by inserting them into the septum on one side of the stopcock. Vacuum from an aspirator is used to pull solvents from squeeze bottles through the needles. Syringe barrels may be attached to the large bore needle on the other side of the stopcock. Solvents from squeeze bottles are introduced into the barrel and removed by the aspirator vacuum. Syringes and needles

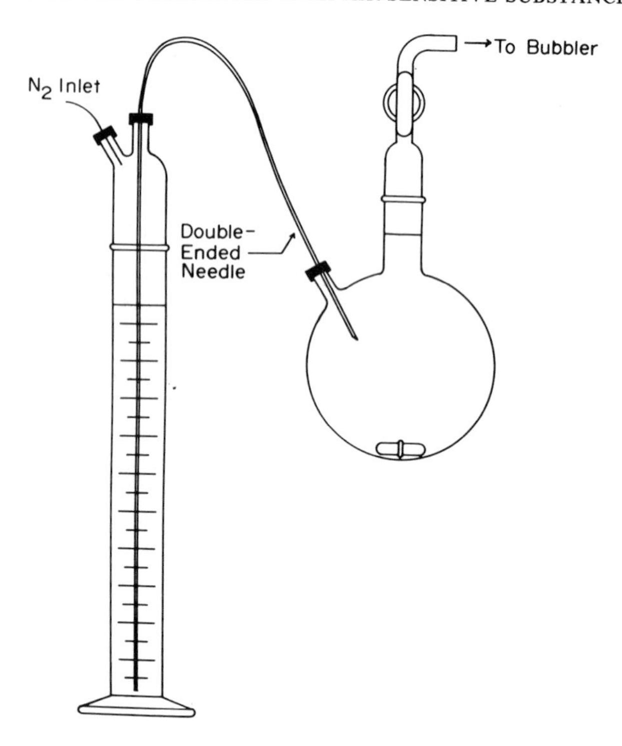

Figure 9.17. Transferring a measured volume of liquid by double-ended needle.

should be dried like ordinary glassware. However, metal-tipped syringes and needles should not be placed directly above the heater in the oven because the localized high temperature may damage the chrome plating.

Double-Ended Needle Techniques

While hypodermic syringes are convenient for transferring small amounts of reagents, larger quantities are not easily handled because larger syringes are difficult to manipulate, multiple transfers with smaller syringes tend to core septa, and plunger freezing is more of a problem if the syringe is used several times. An elegant alternative to multiple syringe transfers has been presented by Shriver.[8] This has become known as the double-ended needle technique **(F9.17)**. A length of stainless-steel needle tubing (~15 in.) is pointed to a standard bevel on one end and to a flat-cut on the other. Alternatively two needles may be connected with a Luer-Lok to Luer-Lok adapter. The standard

[8]Ref. 1, p. 156.

beveled point is inserted through the septum on the storage flask leaving the tip *above* the liquid level; the needle is purged by blowing nitrogen through the storage flask. The flat-cut point is inserted through a prepunctured septum on the measuring vessel which is vented through a bubbler. The needle is placed below the liquid level in the storage flask, and nitrogen pressure is allowed to force the liquid through the needle. When the correct amount of liquid has been transferred, the needle is withdrawn from the storage flask and inserted into the receiver. The bubbler vent is removed from the measuring vessel and connected to the receiver. Gas is blown into the measuring vessel. The resulting pressure differential causes the liquid to be transferred into the receiver. The flat-cut point allows the last few drops of liquid to be transferred. Graduated cylinders or centrifuge tubes with standard taper joints are excellent measuring vessels when fitted with a two-inlet adapter shown in **F9.2A,C,D**. This method is especially good for very sensitive materials which might be difficult to transfer by syringe. It is admirably suited for making standard solutions where the reagent is first placed in a septum-capped volumetric flask and then diluted with solvent via double-ended needle. [This technique can be used in conjunction with various types of filters to remove solids from the liquid. These techniques are mentioned elsewhere in this chapter.] When very large amounts of liquid must be transferred, large-bore needles connected with Teflon tubing are available as Flex-needles from Aldrich-Boranes.

9.3. TRANSFER OF GASES

The organic chemist is often confronted with the problem of quantitatively transferring gases. Solutions to this problem have typically involved vacuum lines, pressure bombs, or gas manifolds. However, such apparatus are often bulky and difficult to handle. They often do not allow the chemist to follow easily the reaction of a gas. Several very simple techniques and apparatus which have been developed to solve these problems are discussed here.

Lecture Bottles

A large variety of gases are available in Lecture Bottles (from Matheson) which usually contain 1 lb or less of gas. These are excellent sources for small quantities of gas. The gas may be removed from the Lecture Bottle by a variety of methods. (For removal of pyrophoric materials from Lecture Bottles, see **S9.9**.) The simplest involves attaching a flexible tube fitted with a syringe needle to the cylinder valve, as in **F9.18**. A bubbler, A, is placed in the line to serve as a safety device. Enough mercury is placed in the bubbler so that the gas will flow through the needle unless the needle becomes plugged. The tube and bubbler

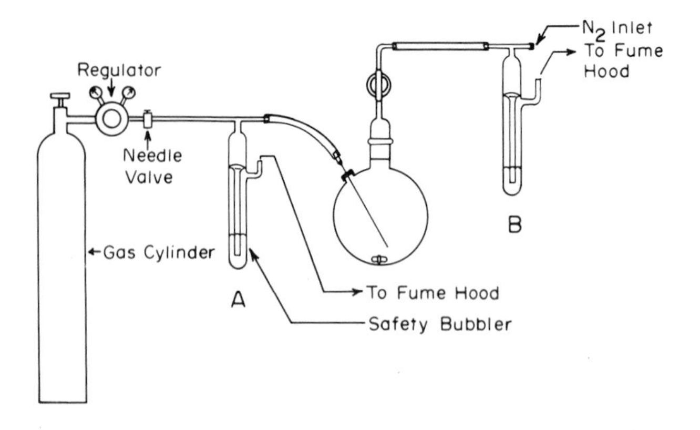

Figure 9.18. Dispensing a gas from a cylinder.

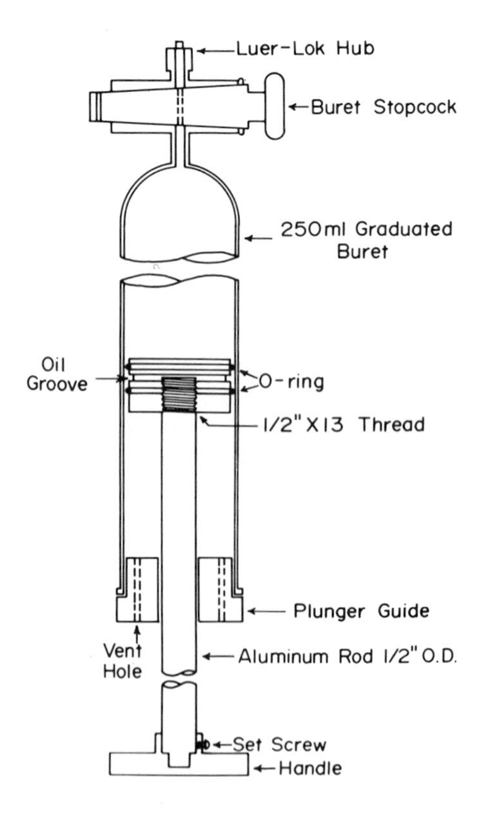

Figure 9.19. Gas syringe.

are flushed with the gas. The needle may be closed when not in use by inserting it into a solid rubber stopper. Prior to use, the cylinder is weighed. If the cylinder is too heavy for weighing, the gas flow rate may be measured with a gas buret (**S9.10**) and then gas added for a known time. The needle is placed through the rubber septum of the reaction flask and the tank valve opened just enough to give a slow stream of gas. The needle is usually placed below the surface of the solvent and the valve opened further. The flow may be monitored by watching the bubbles escaping from the needle. The oil or mercury bubbler, B, will indicate if gas is escaping from the solution. During the addition the bottle may be placed on a balance next to the reaction flask and the weight loss monitored. A 10% excess of gas is usually added to compensate for the inaccuracies of this method (**P2.7**).

When the correct amount of gas has been added, a slow stream of nitrogen is added through the bubbler tee, B. This keeps a positive pressure on the system and prevents the bubbler liquid from being sucked into the reaction flask should a momentary pressure reversal occur, yet does not sweep away gas from above the solution. The needle is first removed from the solution and then the tank valve is closed. After removing the needle from the flask, the cylinder may be reweighed.

This method works well if the gas reacts rapidly or is easily dissolved, and an excess of gas is no problem. In many cases, such as in the hydroboration of ethylene or butene (**P2.7**), the excess gas may be removed after the reaction is over by gently heating the solution while passing a slow stream of nitrogen over it.

Gas Syringes

Developmental reactions are often run on a 5- or 10-mmole scale. These reactions are conveniently followed and analyzed by chromatography and give enough product for easy isolation and characterization of the products. Such reactions require the accurate addition of relatively small amounts of gas. Large gas-tight syringes are available from the Hamilton Co. Gas-tight syringes may also be constructed from readily available laboratory equipment (**F9.19**).[9] A 250-ml buret is convenient for the syringe barrel, but larger or smaller syringes may be constructed.

A few drops of silicone or mineral oil are placed in the oil groove prior to use. The needle is placed in a rubber stopper and the plunger depressed to leak test the syringe. It should be possible to compress the air to half its original volume with no evidence of a leak.

The gas syringe is filled as shown in **F9.20**. The cylinder is opened to flush the system with gas. The flow may be monitored with the mercury

[9]G. W. Kramer, *J. Chem. Ed.*, **50**, 227 (1973).

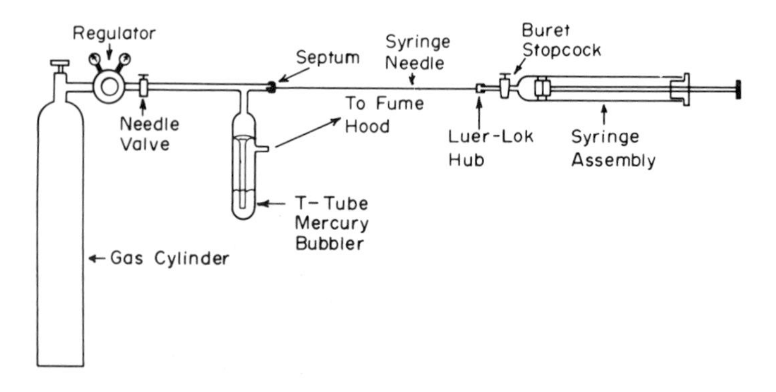

Figure 9.20. Filling the gas syringe.

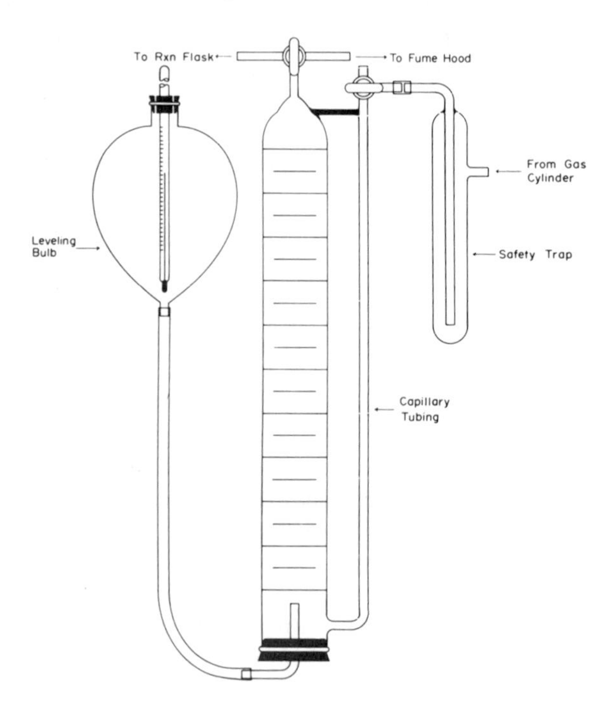

Figure 9.21. Gas-transfer buret.

bubbler. For gases under only moderate pressure, such as the butenes, the cylinder valve may be capped directly with a wired-on septum. The syringe is filled with gas, then emptied into a fume hood several times to purge air from the syringe. The correct volume of gas, calculated from the gas laws, is drawn into the syringe and the stopcock closed while the needle is still in the septum. Just before use, the stopcock is opened, and the excess pressure in the syringe gives the needle a final purge.

The reaction flask is adjusted to atmospheric pressure using the mercury bubbler as a gauge. (When the mercury in the inner and outer tubes of the bubbler is at the same level, the flask is at atmospheric pressure.) The syringe needle is then inserted through the septum of the reaction flask and the gas slowly injected. The mercury bubbler should not bubble during this time. The syringe needle is left in the flask until the bubbler again indicates atmospheric pressure.

This syringe technique may be repeated several times for larger scale reactions. We have successfully used the syringe for the addition of butenes and ethylene to hydroboration mixtures, addition of gaseous polyhalomethanes,[10] and the addition of gaseous alkynes to *n*-butyllithium.

Gas Buret

Known quantities of gas may be transferred by using a gas buret (**F9.21**). The gas buret is constructed from common laboratory equipment. A syringe needle is placed on the tube leading to the reaction flask. The needle may be connected to a bubbler leading to a fume hood prior to transferring the gas to the reaction flask.

The buret is filled with mercury, mineral oil or dibutyl phthalate to avoid saturating the gas with water. The system is flushed several times with the gas by opening the stopcock to the gas cylinder and buret, and closing the top stopcock. To facilitate filling the buret the leveling bulb may be lowered as the gas enters the buret. To empty the buret, the stopcock to the cylinder is closed and the top stopcock opened to the fume hood or reaction flask tube. The leveling bulb is then raised, and the liquid pumps the gas out of the buret. To measure the volume of gas, the level of liquid in the buret and bulb must be equal. After the correct volume of gas is placed in the buret, the needle is removed from the bubbler and placed through the septum inlet of the reaction flask. The process may be repeated several times to pump a large volume of gas into the reaction flask.

Automatic Gasimeter

Many noncondensable gases may be generated and transferred using the automatic gasimeter (**S9.4**).

Condensable Gas Trap

Gases which have fairly high boiling points, such as butadiene, sulfur dioxide,

[10] H. C. Brown, B. A. Carlson, and R. H. Prager, *J. Amer. Chem. Soc.*, 93, 2070 (1971).

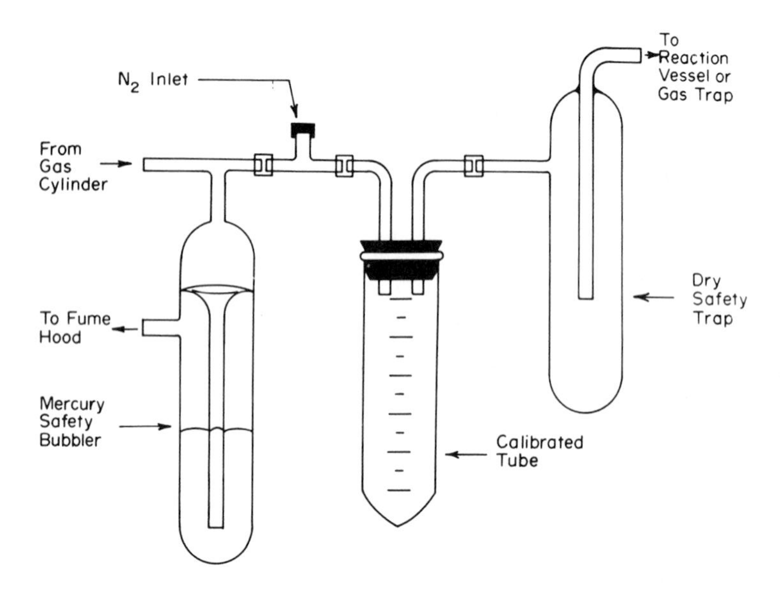

Figure 9.22. Simple apparatus for transferring condensable gases.

or boron trichloride, may be condensed in a low temperature bath and measured as the liquid. A simple gas trap may be constructed as shown in **F9.22**. The assembled apparatus is flushed with nitrogen with a needle connected into the tee on the bubbler. A slow stream of nitrogen is passed through the tee on the bubbler and the graduated tube cooled low enough to condense the gas. The cylinder is opened and the gas condensed until the correct volume has been collected.[11] Alternatively, the tube may be quickly disconnected, capped, and weighed to determine the amount of gas collected. The nitrogen needle is then removed from the tee and quickly placed through the septum inlet of the reaction flask. The gas is then distilled into the reaction vessel through the dry safety trap by removing the cold bath.

The above trap serves its purpose for the worker who only occasionally uses condensable gases. For use with corrosive gases, a more elegant, all-glass apparatus is shown in **F9.23**. The system may be oven dried, assembled hot, and flushed with nitrogen. The three-way stopcock is opened to a suitable gas trap and fume hood. The centrifuge tube is cooled in a Dry Ice-acetone bath and the correct volume of gas condensed. A slow stream of nitrogen is admitted and then

[11]Liquified gas densities are available in the *International Critical Tables*, McGraw-Hill, New York, 1926, or *Physical Constants of Hydrocarbons*, American Chemical Society Monograph Series No. 18, Reinhold Publishing Corp., New York, 1939.

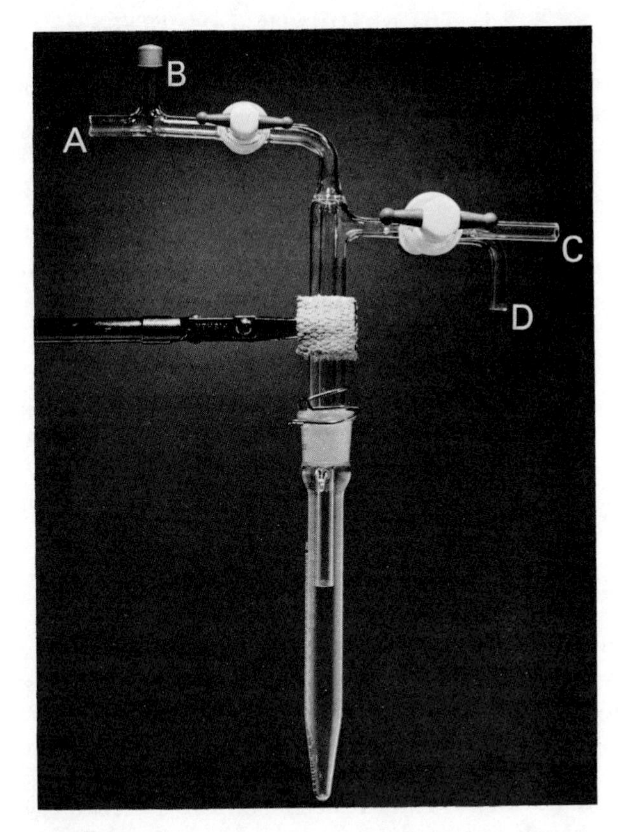

Figure 9.23. All-glass apparatus for transferring condensable gases: (A) to gas cylinder; (B) nitrogen inlet; (C) to reaction flask; (D) to gas trap and hood.

the tank valve closed. The three-way stopcock is opened to the reaction flask and the gas allowed to distill into the system. The gas is never exposed to the air in this system.

Solutions of Gases

It is sometimes convenient to store and transfer small quantities of gases as standard solutions. For such storage, the gas must be fairly soluble, and a convenient method of analysis must be available. Examples of this method are borane in THF, boron trichloride in ethyl ether (**P4.12**) and boron trichloride in pentane. Such solutions are fairly stable when stored at approximately 0°. However, since gas may escape from solution or react with the solvent, the solutions should be analyzed prior to use.

9.4. THE AUTOMATIC GASIMETER

The automatic gasimeter[12] (**F9.24**) was originally developed to overcome problems associated with hydrogenations. The apparatus is readily adapted for the production of a variety of gases. Several are listed in **T9.1**. Other gases may

Table 9.1. Gases Generated by the Automatic Gasimeter

Gas	Liquid in Buret	Material in Generator	Ref.
H_2	$NaBH_4$ in ethanol or water	Conc. HCl or CH_3CO_2H	12
CO	HCO_2H	Conc. H_2SO_4	13
HCl	Conc. HCl	Conc. H_2SO_4	14
HBr	Conc. HBr	PBr_3/ethyl acetate	15
O_2	Aq. H_2O_2	MnO_2/H_2O, OH^-	16
$H_2C{=}CH_2$	1,2-dibromoethane	Zinc	17
SO_2	Aq. $NaHSO_3$	50% H_2SO_4	18

be generated as needed, provided that the following basic requirements are met. The gas must be generated from the reaction of the liquid (or solution of a reactant) in the buret with a liquid or suspension in the flask. The reaction must be rapid and preferably quantitative. If it is not quantitative, it must be reproducible. No other gaseous products which cannot be removed by appropriate traps can be formed. The gas should not react appreciably with mercury, and it should be relatively insoluble in the reaction solvent. This latter problem can be overcome if the solvent is saturated with the gas before the substrate is added.

The apparatus consists of a buret attached to a 4-in. needle which is placed through a septum into the automatic valve. The valve consists of a drip-tip tube (which must be thoroughly cleaned to give good dripping) with several holes near the top which is filled with mercury. The depth of the needle in the mercury may be adjusted so that the height of the mercury equals the pressure of the liquid in the buret. The liquid in the buret, interacting with the material in the generator, produces the gas.

The apparatus is equipped with septum inlets so that solutions may be added by syringe. The reactant in the reactor flask absorbs gas when the stirrer is

[12]C. A. Brown and H. C. Brown, *J. Org. Chem.*, **31**, 3989 (1966). The automatic hydrogenator available from Delmar Scientific Laboratories, Inc., 317 Madison Avenue, Maywood, Illinois, was utilized for the applications. An improved model, specially designed for applications involving the automatic gasimeter is now available from the Ace Glass Co., Vineland, New Jersey.

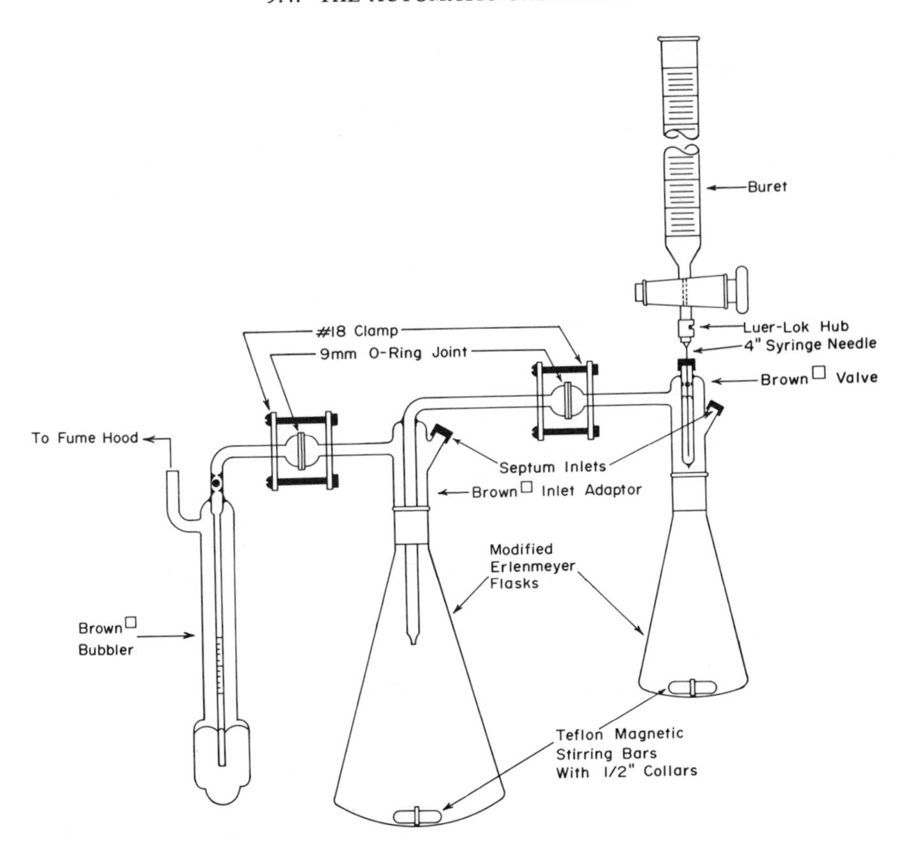

Figure 9.24. Automatic gasimeter.

started and the internal pressure of the system drops 20-30 mm. The combined pressure of the system and the height of mercury in the valve then no longer support the liquid in the buret. The liquid flows through the needle and drips into the material in the generator. The gas produced brings the system back toward atmospheric pressure, and the flow of liquid stops until more gas is absorbed. By using a standardized liquid in the buret, the course of the reaction may be followed to within ±1% accuracy.

The mercury bubbler seals the system from the atmosphere and serves as a pressure-release safety valve. A ball valve at the top of the bubbler prevents mercury from being pulled into the system should the reaction absorb gas too fast or the buret needle become plugged. The mercury bubbler also serves as a sensitive monitor for progress of the reaction since the mercury level rises and falls as the reaction proceeds.

High-speed stirring is essential if one is following the rate of a reaction. In **F9.24** special Erlenmeyer flasks with rounded bottoms are shown. Such flasks provide superior stirring capabilities over round-bottom flasks or regular Erlenmeyer flasks. With round-bottom flasks, the stirring bar often gets off center and may fly through the wall of the flask during high-speed stirring. High-speed stirring is further facilitated by using Teflon collars on the stirring bars.

The apparatus may be altered to fit the need of the reaction. Gas purifiers may be placed directly in the tube between the generator and the reactor or in U-shaped tubes. Such U-tubes may be placed in a cold trap or a condenser may be placed between the generator flask and the remainder of the system to serve as a solvent trap.

In addition to the advantages previously described, the automatic gasimeter has no serious limitations in capacity for laboratory scale reactions. A fraction of a mmole to as much as 2000 g of material has been hydrogenated with no difficulty.

For organoborane work, the production of carbon monoxide (**P8.3**, **P8.5**) and oxygen (**P6.12**) are particularly important and will be discussed in detail.

Carbon Monoxide[13]

Carbon monoxide may be generated by the dehydration of formic acid.

$$HCO_2H \xrightarrow{H_2SO_4} H_2O + CO$$

The gas is entirely contained inside the system so that reactions may be run on the bench top, provided that the mercury bubbler exit is connected to a well-ventilated hood.

The buret is filled with anhydrous formic acid and the generator with about 100 ml of concentrated sulfuric acid heated to 90-100°. Ascarite is placed in the tube between the generator and the reactor to absorb carbon dioxide. Care must be taken that this tube does not become plugged. The Ascarite should be replaced between reactions if it has turned white. The organoborane may be prepared or placed in the reactor in an appropriate solvent under nitrogen. Then the system is flushed with carbon monoxide by slowly injecting 2 ml of formic acid into the generator. After completion of the flushing process, enough carbon monoxide is withdrawn by syringe to bring the system to atmospheric pressure as indicated by the mercury bubbler. Stirring initiates the absorption of carbon monoxide. The reaction may be followed by reading the buret. One

[13] M. W. Rathke and H. C. Brown, *J. Amer. Chem. Soc.*, 88, 2606 (1966).
[14] H. C. Brown and M.-H. Rei, *J. Org. Chem.*, **31**, 1090 (1966).
[15] Research with M.-H. Rei.

milliliter of formic acid generates 26.6 mmoles of gas. The carbonylation of organoboranes provides a route to tertiary alcohols (**P8.3**), ketones (**P8.4**), and aldehydes (**P8.5**).

Oxygen[16]

Oxygen may be produced by the catalytic decomposition of hydrogen peroxide.

$$2\,H_2O_2 \xrightarrow[MnO_2]{cat.} 2\,H_2O + O_2$$

The buret is filled with standardized aqueous hydrogen peroxide (3-30%). The hydrogen peroxide may be standardized using a gas buret (**S9.10**) in the following manner. The gas buret hydrolysis flask is charged with 3 g of manganese dioxide (technical grade), 5 ml of water, and 2 ml of 3*M* sodium hydroxide. The flask is immersed in a water bath. An aliquot of hydrogen peroxide solution (15 ml of a 3% solution generates 220 ml of oxygen) is injected into the flask and the oxygen measured. The concentration of the solution is then calculated using the hydride molarity equation (**S9.10**) with the exception that the volume displaced by the aliquot (V_B) is not subtracted from the volume of the oxygen (V_{H_2}). The liquid displaces an equal volume of gas in the gasimeter so that it adds to the concentration of the solution.

The gas generator flask of the gasimeter is charged with 3 g of manganese dioxide (technical grade), 5 ml of water, and 2 ml of 3*M* sodium hydroxide, then immersed in a room-temperature water bath. The tube leading from the generator is filled with glass wool. The tube leading to the reactor is filled with sodium hydroxide or potassium hydroxide pellets to absorb moisture. Since organoborane solutions rapidly absorb oxygen, even without stirring, a special oxygen flushing technique is used. A 100-ml flask is placed on the reactor portion and the system purged with oxygen by injecting 15 ml of 30% hydrogen peroxide into the generator. The organoborane is placed or prepared in a separate flask in a solvent under nitrogen. The 100-ml flask is removed from the apparatus and quickly replaced with the flask containing the organoborane. The remaining nitrogen above the solution is then removed by carefully, but quickly, injecting 2-3 ml of 30% hydrogen peroxide into the generator. This process minimizes the amount of oxygen absorbed by the organoborane solution during the flushing process. Initiation of stirring causes a rapid absorption of oxygen. The course of the oxidation may be followed by reading the buret. For oxidation of the organoborane to the alcohol, the oxidation may be stopped when exactly 1.5 moles of oxygen per mole of borane has been absorbed. The reaction of organoboranes with elementary oxygen provides a route to alcohols (**S5.9**) and alkyl hydroperoxides (**P6.12**).

[16] H. C. Brown, M. M. Midland, and G. W. Kabalka, *J. Amer. Chem. Soc.*, 93, 1024 (1971).

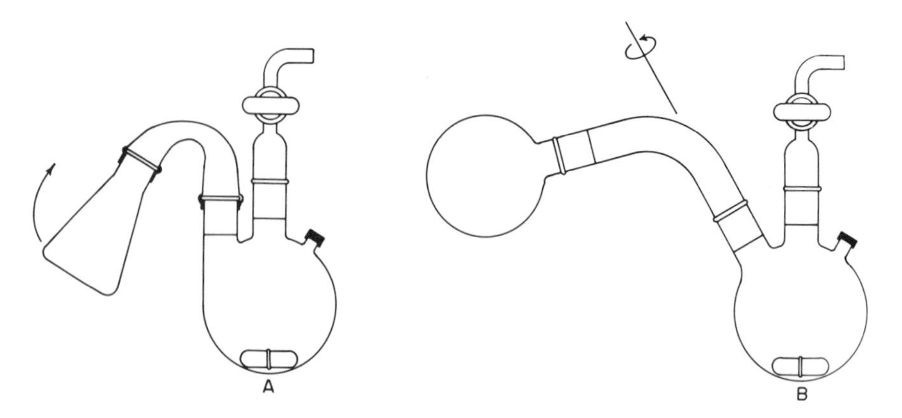

Figure 9.25. Addition of solids: (A) utilizing an Erlenmeyer flask and rubber tubing; (B) utilizing a 105° gooseneck and round-bottom flask with male joint.

9.5. TRANSFER OF SOLIDS

Inert atmosphere manipulation of solids is often a difficult problem. The surest solution involves the use of an inert atmosphere glove box. The simplicity of this solution is offset by the cost of the apparatus and the problems involved in maintaining a workable system. A cheaper alternative is the plastic glove bag, manufactured by Instruments for Industry and Research, or ordinary plastic bags. Although these enclosures are satisfactory for moderately sensitive materials, they are somewhat cumbersome to use, and it is difficult to obtain and sustain a very inert atmosphere in them.[19] Simple operations, like opening and capping reagent bottles or other simple transfers, can be carried out in these glove bags.

There is no foolproof method for adding a solid to an air-sensitive reaction mixture. If at all possible, solid materials should be used in solution, where they can be handled using liquid transfer techniques. Where this is not possible, some thought should be given to transferring them as slurries or suspensions. Slurries of finely divided solids can be transferred, like liquids, using large-bore (15-17 gauge) needles.

If a dry solid must be added to a air-sensitive reaction mixture, it can be done using the apparatus shown in **F9.25A (P6.18)**.[20] The solid is loaded into the

[17] Private communication from C. A. Brown, Cornell University, Ithaca, N.Y.
[18] Research with P. Jacob, III.
[19] Descriptions of technique used with inert atmosphere enclosures can be found in Ref. 1, Chapter 8. These methods will not be repeated here.
[20] L. F. Fieser and M. Fieser, *Reagents for Organic Synthesis,* Vol. I, John Wiley and Sons, Inc., New York, 1967, p. 24.

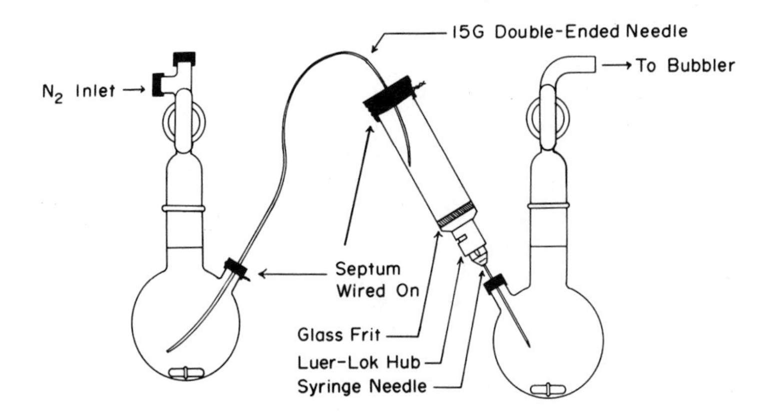

Figure 9.26. Transferring a liquid through an in-line filter.

addition flask A. If the solid itself is air-sensitive, this loading must be done in an inert atmosphere enclosure. The loaded addition flask is flushed with nitrogen. This is best done using the evacuation/refill method **(S9.1)**. The addition flask is attached to the reaction flask with a length of large diameter rubber tubing (Gooch tubing) while maintaining a flow of nitrogen through the reaction flask. The solid is added to the reaction mixture by inverting the addition flask. **F9.25B** shows a slightly more elegant setup. The addition tube is connected to the reaction flask with a 105° gooseneck adapter. The solid is added by rotating the addition-tube assembly upward **(P8.8)**. To minimize contamination of the solid with stopcock grease, the addition flask should have a male standard-taper joint.

9.6. FILTRATION OF REACTION SOLUTIONS

Solutions containing suspended impurities or precipitates which must be removed or isolated are easily filtered under nitrogen through sintered-glass filters. For small samples or solutions containing only a small amount of suspended solids **(P4.7, P4.8)**, or when a large filter area is not required, a simple filter may be used **(F9.26)**.[21] The filter is constructed by joining a large Luer glass joint to a 10-mm coarse, sintered-glass sealing tube. A metal Luer-Lok is attached to the joint **(S9.2)**. The other end of the glass tube is rolled slightly so that a rubber serum cap may be wired securely in place. A needle is attached to the dry apparatus and the system flushed with nitrogen. The needle is then inserted through the septum-capped inlet of a receiving flask. The solution to

[21] Commercial filters of this type are available from Aldrich Chemical Co.

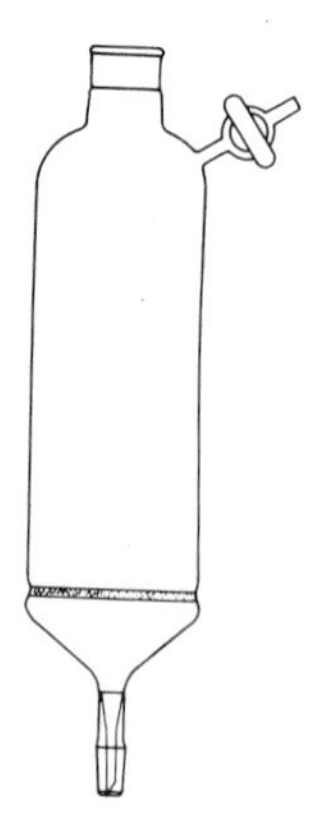

Figure 9.27. Filter chamber. This device is useful for removal of large amounts of solid.

be filtered is transferred into the filter via a large-bore, double-ended needle. A positive pressure of nitrogen facilitates the filtration.

For most purposes, a coarse filter disk is adequate. The finer disks are inconveniently slow and tend to become plugged. Even when using a coarse disk it is advisable to have several filters available in case the frit becomes plugged.

For larger amounts of liquid, when a large amount of solid is to be collected or separated **(P4.12)**, or when a large filter area is required [such as in the filtration of lithium aluminum hydride solutions **(S9.11)**], a large filter chamber[22] may be used **(F9.27)**. The apparatus may be constructed by sealing a glass tube of similar diameter onto a coarse filter funnel which has a standard taper joint. The tube is closed at the top with a ground-glass stopper such that the volume enclosed is 250-500 ml. A 2-mm stopcock with 9-mm medium-walled tubing is attached near the top.

The filter is dried in an oven and then flushed with nitrogen with a receiver flask attached. The receiver flask may be attached to a mercury bubbler via a sidearm or by a needle placed through the rubber septum. The solution is transferred into the filter via a double-ended needle through the stopcock and filtered under a positive nitrogen pressure. If the solid is to be isolated, it may be removed through the top in a glove bag.

When filtering a hydride solution, it is extremely important that the sintered-glass disk be dry, as moisture will hydrolyze the hydride and deposit salts in the disk. This will quickly plug the filter. A 2-in. layer of a filter aid, such as Celite (Johns-Manville), should be used for hydride solutions. The Celite must be dried in an oven at 125° for 12 hours prior to use.

[22]Commercial filter chambers of this type are available from Alfa Product Division of Ventron Corporation.

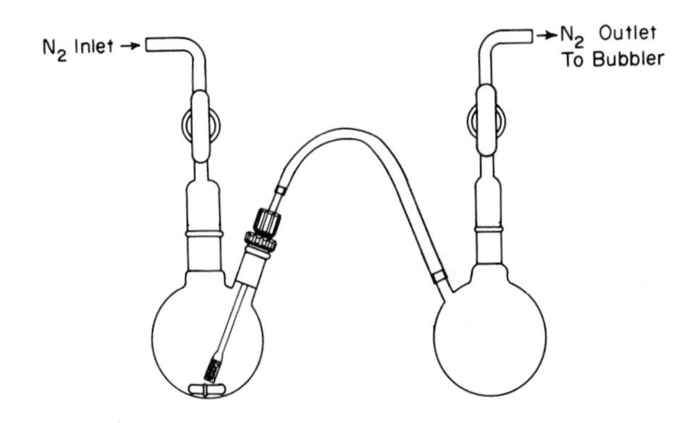

Figure 9.28. Filtration through a gas-dispersion tube.

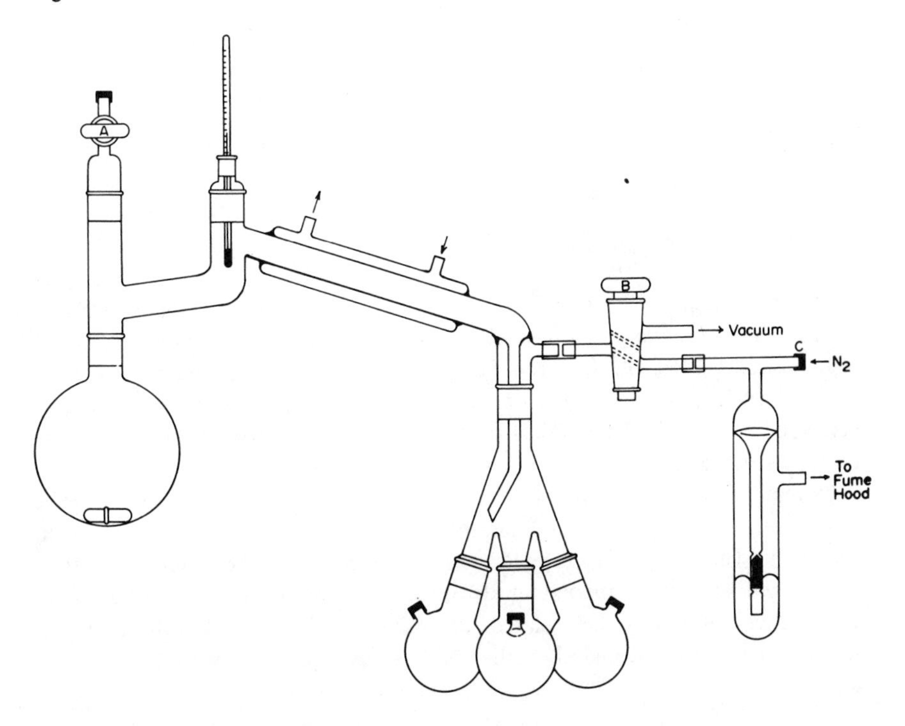

Figure 9.29. Basic distillation apparatus for air-sensitive materials.

Another filter device is shown in **F9.28**. This device consists of a dry, coarse gas-dispersion tube immersed in the liquid and connected to the receiving flask

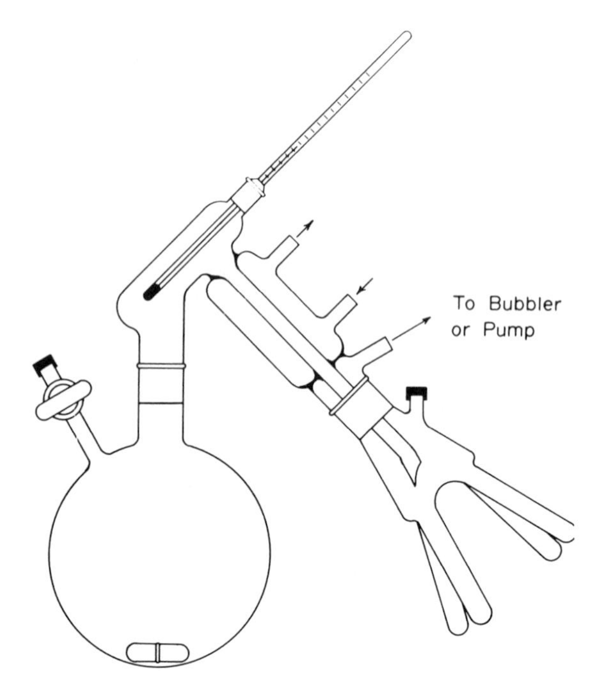

Figure 9.30. Micro-distillation apparatus.

by flexible tubing. The liquid is forced into the receiving flask by nitrogen pressure. If the liquid is the desired component, such as in the preparation and isolation of 9-BBN (**P2.11**), the flexible tubing should be made of an inert material such as Teflon. This device has an advantage over those previously described in that the filter flask may be heated by an oil bath or heating mantle during the filtration.

To remove a large amount of a finely divided precipitate, the solution may be transferred via a large-bore double-ended needle into a dry, nitrogen-flushed centrifuge tube which is capped with a serum cap. The solution is then centrifuged and the supernatant decanted via a double-ended needle. If an air-sensitive solid is to be isolated, the solid may be washed with additional solvent, recentrifuged, and then the solid removed in a glove bag.

9.7. DISTILLATION OF AIR-SENSITIVE PRODUCTS

The techniques involved in distillation are familiar and have been comprehensively discussed elsewhere.[2,3] Standard glassware with appropriately positioned septum inlets is usually adequate for air-sensitive materials (**F9.29-31**).

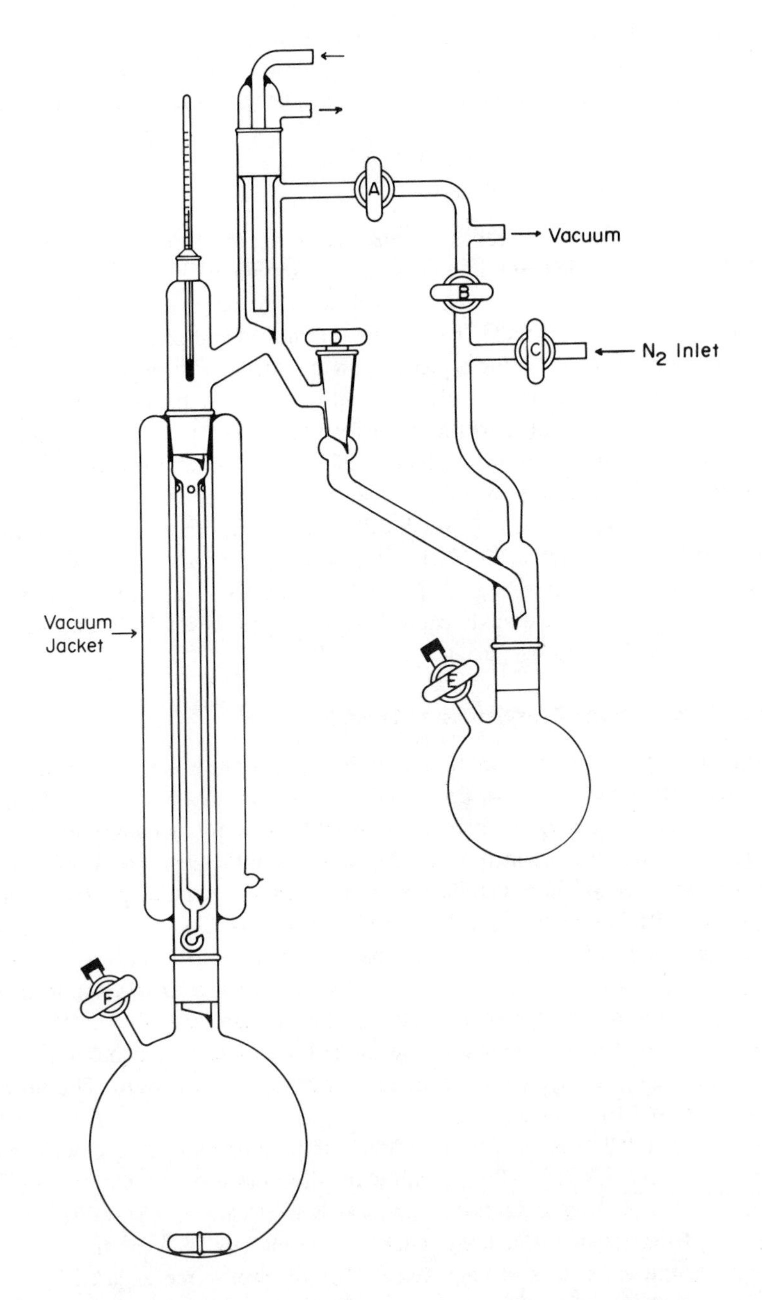

Figure 9.31. Fractional distillation of an air-sensitive material using a micro Widmer column (Lab Glass No 5930-S).

It is useful to have a Teflon stopcock-controlled septum inlet on the pot since hot vapors from reactive materials may deteriorate the septum (**F9.30**). Alternatively, a round-bottom flask and Claisen head (**F9.29**) with adapter (**F9.2**) may be used.

Distillation of Air-Sensitive Material

Simple distillation of air-sensitive material (e.g., cyclopentyldichloroborane, **P4.13**) may be carried out in the apparatus shown in **F9.29**. The oven-dried glassware is assembled hot and flushed with nitrogen through stopcock A, preferably using a long needle. The needle is removed and placed through septum C on the bubbler maintaining a gentle flow of nitrogen throughout the distillation. All joints should be securely fastened with rubber bands or spring clips.[4] The organoborane is added to the pot via syringe (**S9.2**). Stopcock A is closed, and the distillation carried out in the normal manner. Fractions may be collected by realigning the receivers. When the distillation is complete, the pot is allowed to cool with nitrogen flowing through the bubbler. The fractions are removed for storage (**S9.9**) or analysis (**S9.10**) by standard syringe techniques. If flasks with sidearms are not available, regular round-bottom flasks may be used. The whole receiver is removed and quickly capped with a large septum. Individual fractions may be removed by syringe, using a long needle.

Vacuum Distillation of Air-Sensitive Materials

Distillation of air-sensitive materials under vacuum can be done in common apparatus. Obviously, a provision for adding material to the pot must be made. Since the apparatus must be bled down with nitrogen, it is convenient to use a parallel-arm three-way stopcock (**F9.29**) with a bubbler and check valve in the nitrogen line. A vacuum distillation of an organoborane (e.g., tri-secondary butylborane **P2.7**) may be carried out in this apparatus. The oven-dried glassware is assembled hot and flushed with nitrogen. The pot is charged via syringe and the system opened to vacuum. If a bleed is required, a nitrogen line may be attached to an appropriate needle valve in the vacuum system.

After the distillation is complete and the pot has cooled, stopcock B is opened to nitrogen, and the system is returned to atmospheric pressure. The products can be removed by syringe.

Fractional distillation of closely boiling compounds requries a more sophisticated apparatus (**F9.31**). This assembly requires more manipulation to collect fractions. When cutting a fraction, stopcocks B and D are closed, and the receiver is opened to nitrogen through stopcock C. The material may be transferred via syringe techniques into a storage vessel. The receiver is reevacuated by closing stopcocks A and C and opening B. Then stopcock A is opened and the distillation continued. When desirable, the receiver may be changed by first opening it to

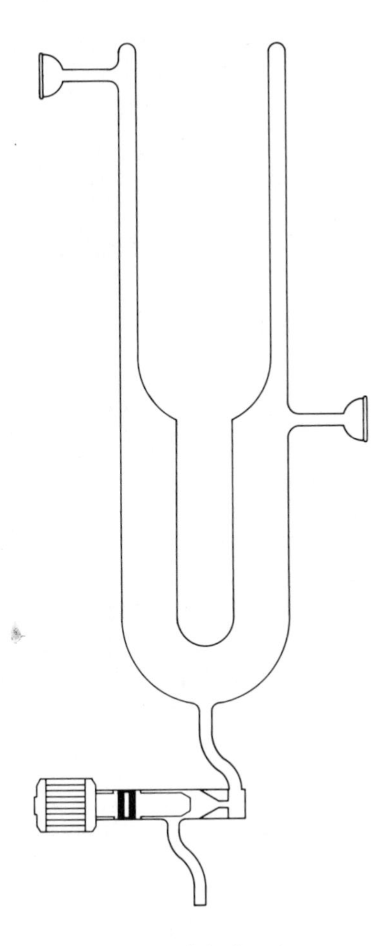

Figure 9.32. Solvent catcher. This trap is useful when large amounts of solvent must be removed.

nitrogen, as described. With nitrogen flowing through a needle inserted through stopcock E, the flask is removed and quickly capped with a connecting tube. A fresh flask is connected to the still. It is then reevacuated as previously described, and the distillation is continued.

Solvent Stripping from Air-Sensitive Materials

Concentrating the product from larger amounts of solvent prior to distillation is a more acute problem for air-sensitive materials since rotary evaporators are not convenient for inert-atmosphere work. If the solvent and product have sufficiently different vapor pressures, it may be possible to evacuate the flask

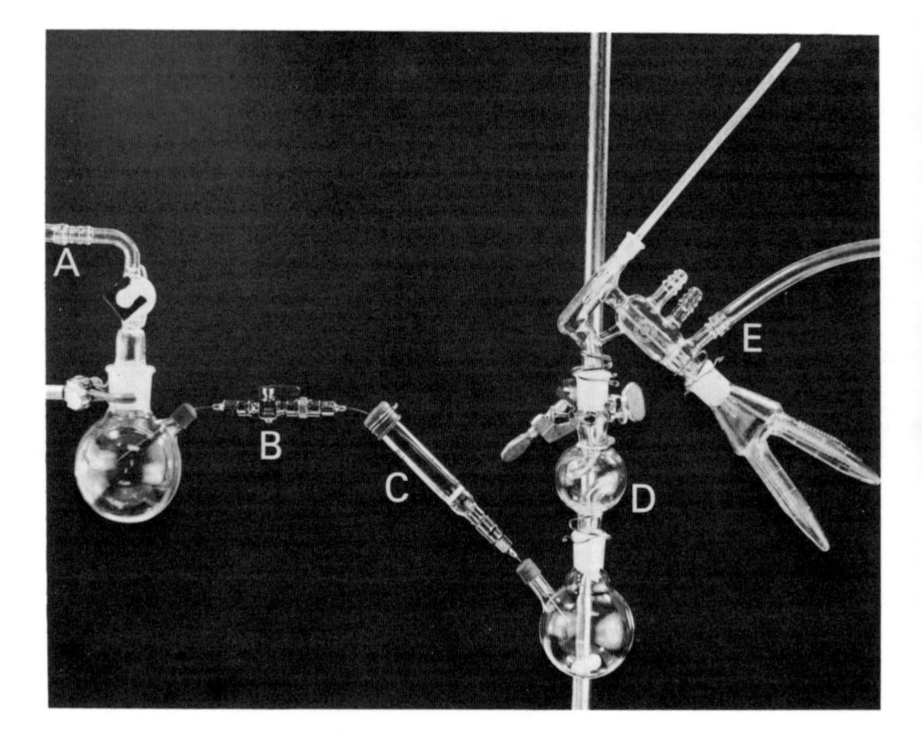

Figure 9.33. Solvent stripping from air-sensitive materials: (A) nitrogen inlet; (B) valve with two Luer-Loks (B-D No. 3155 or No. 3152 and adapter No. 3114); (C) in-line filter; (D) splash guard; (E) to solvent catcher and vacuum.

with an aspirator, remove the solvent, and transfer the residue by syringe. This technique has several disadvantages. It is generally difficult to avoid foaming, and furthermore, removal of a solvent, such as THF, is generally slow through a connecting tube. Significant amounts of product are also usually lost in small-scale reactions when the pure material is withdrawn by syringe. These difficulties may be surmounted in an apparatus in which adapter A, **F9.29**, has been replaced with a cylindrical separatory funnel. The solution is added to the funnel by syringe. The system is connected through a parallel-arm three-way stopcock to a vacuum source and nitrogen, and the apparatus is evacuated. The solution is then added at such a rate that the solvent flash evaporates. If an oil pump is used to provide the vacuum, an efficient solvent-catching trap should be used (**F9.32**) in the vacuum line. Where solids must be removed, a fritted filter can be used to remove suspended material from the solution before addition to the dropping funnel. Alternatively, an apparatus such as that shown in **F9.33** is convenient. The crucial item is

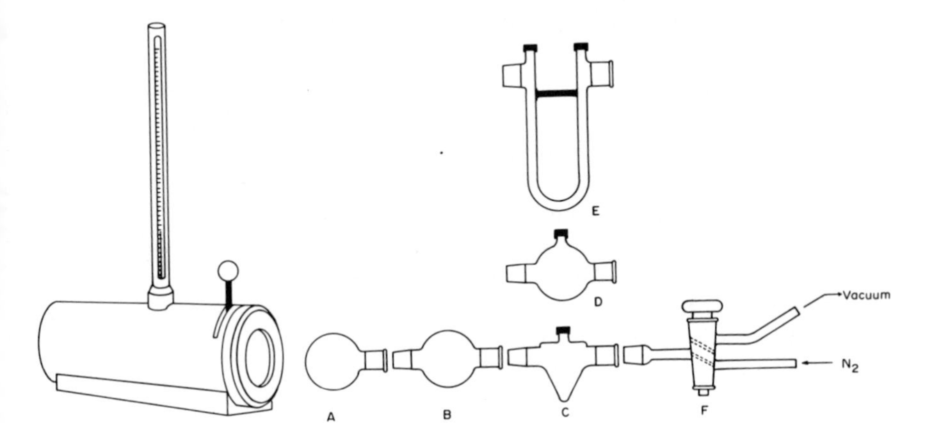

Figure 9.34. Kügelrohr distillation apparatus.

a valve with two Luer-Loks **(S9.2)**. When the system is opened to vacuum, the solution is pulled into the distilling pot, and the solvent is flash-distilled into the solvent trap. If solvent begins to build up in the pot, valve A can be temporarily closed. A splash guard in the system prevents foam-overs during solvent removal. The remaining material is distilled and isolated in the usual manner.

Bulb-to-Bulb Distillation of Air-Sensitive Materials

Distillation of semi-micro scale reaction products in normal equipment leads to significant losses of material. Where fractionation is unnecessary, bulb-to-bulb distillation provides an attractive alternative. Kügelrohr ovens and associated glassware provide a simple, bench-top means of carrying out bulb-to-bulb distillations **(F9.34)**. To distill small amounts of material in this apparatus, flasks A and B are dried in an oven, assembled hot, temporarily capped with a septum, and flushed with nitrogen. The pot is charged by syringe. The receiver (C, D, or E) is attached to the stopcock while hot and flushed with nitrogen through the stopcock. With nitrogen flowing, the temporary septum is removed and the assembly completed. The flasks may be held together with spring clips.[4] If the distillation is to be done under vacuum, the apparatus is held in a vertical position and the stopcock opened to vacuum. When the foaming ceases, flasks A and B are placed in the oven (B serves as a foam trap). After distillation is complete, stopcock F is opened to nitrogen and the system repressurized. The product may be removed by syringe. Quantities ranging from 50 mg to 5 g may be distilled in a very short time using this apparatus.

9.8. RECRYSTALLIZATION AND SUBLIMATION

Recrystallization of sensitive materials can be accomplished in special Schlenk-tube apparatus. Techniques for using this type of glassware are adequately described elsewhere.[23] Alternatively, recrystallization may be carried out using common glassware. The technique is the same as a normal, open recrystallization except that all transfers and separations are done under nitrogen.

The material to be recrystallized is placed in a flask fitted with a septum inlet, a magnetic stirring bar, and a reflux condenser topped with a stopcock-controlled connecting tube. Some solvent is transferred into the flask via syringe or double-ended needle. The recrystallization flask is heated to reflux with stirring. If the solid does not dissolve, more solvent is added. When the solvent is added by double-ended needle, it is convenient to leave the needle in place until the solid dissolves. The amount of solvent added can be controlled by raising and lowering the needle in the solvent reservoir flask. Once the solid is dissolved, the reflux is terminated, and the contents are allowed to cool slowly without stirring. When the solid has crystallized, it is separated from the supernatant liquid by inert atmosphere filtration (**S9.6**) or by the following procedure.

If the solid settles nicely, the liquid may be removed by decantation with a double-ended needle or syringe. If the crystals are small and do not settle well, the suspension is transferred through a large-bore (15-gauge) double-ended needle into a centrifuge tube.[24] The material is centrifuged, and the liquid decanted through a double-ended needle. If a small amount of solid is still suspended after centrifuging, it may be removed from the supernatant liquid with an in-line filter, such as that shown in **F9.26**. The main advantage of this decantation procedure is that precipitates which would clog a fritted disk are easily handled.

After the liquid is removed, the precipitate can be washed with cold solvent by cooling the recrystallization flask in a bath, and then adding precooled solvent by double-ended needle. The cold crystals are stirred with the solvent. After the solid settles, the liquid is removed by decantation. The solid can be dried by evacuating the recrystallization apparatus through the connecting tube. As with unsensitive solids, care must be taken during vacuum drying to prevent sublimation of the solid out of the flask. This can be prevented at the expense of longer drying times by maintaining cold water flow through the condenser. If the solid sublimes, it will be caught by the condenser. If the solid is to be used in solution, the sublimate can be washed off the condenser by adding solvent through the connecting tube inlet.

[23] Chapter 7 of Ref. 1.

[24] Crown-cappable centrifuge tubes are available through Texas Alkyls.

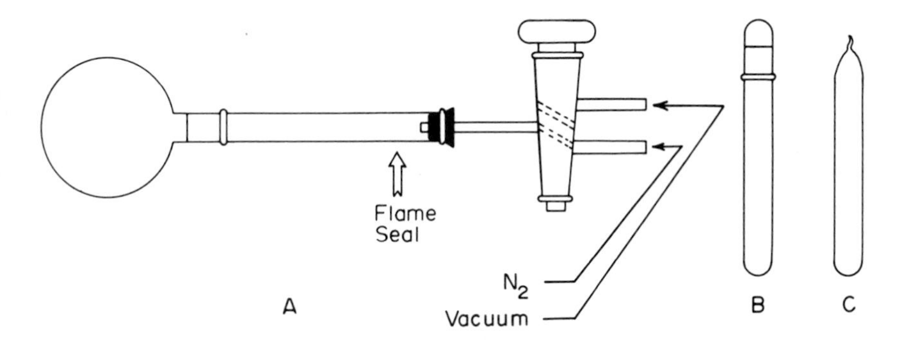

Figure 9.35. Glassware for Kügelrohr sublimations.

A useful alternative to recrystallization is sublimation. Standard sublimation techniques have been covered elsewhere.[2] One disadvantage to sublimation is that compounds which have similar vapor pressures cannot be easily separated. Commercial sublimators, such as the McCarter design, are adequate for air-sensitive compounds.[25] However, they must be opened in an inert-atmosphere box or a glove bag. Transporting the apparatus into such a controlled-atmosphere vessel frequently results in loss of purified material because the product falls back into the pot.

For modest quantities, a more convenient procedure to affect sublimation involves the use of a Kügelrohr oven. The apparatus is pictured in **F9.34**. The dried flasks are placed in a glove bag where the pot is charged with the material to be sublimed. The apparatus is assembled in the glove bag, and the stopcock is closed. The assembly is then moved to the bench and sublimed by inserting the pot into the Kügelrohr oven. If necessary, a vacuum may be placed on the apparatus through one of the stopcock arms. If this is done, the apparatus must be slowly bled down with nitrogen [preferably through a tee-tube bubbler **(F9.7C-E)**] after the sublimation is completed. The purified material is then removed from the receiver tube in a glove bag. If the specialized Kügelrohr glassware is not available, an acceptable alternate assembly may be constructed as shown in **F9.35A**.

If the sublimed material is to be stored, the receiver tube can be flame cut and sealed above the sublimate and the bottom end closed with a standard-taper cap **(F9.35B)**. Use of a male joint on the receiving end of the sublimation apparatus ensures that the material will not become contaminated with stopcock grease. Alternatively, the sublimation can be carried out in a test tube fitted with a rubber stopper and stopcock. After sublimation, the product can be removed for

[25] R. J. McCarter, *Rev. Sci. Inst.*, **33**, 388 (1962).

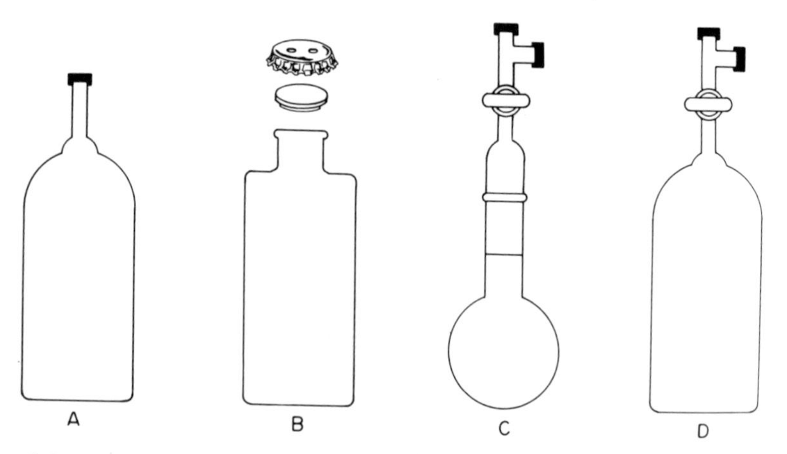

Figure 9.36. Inert-atmosphere storage containers: (A) septum-capped ampule; (B) crown-cappable bottle; (C) round-bottom flask and adapter **(F9.2A)**; (D) stopcock-protected ampule.

storage or a protective ampule **(F9.35C)** formed by making flame seals on the tube.

9.9. STORAGE OF AIR-SENSITIVE MATERIALS

Storage of Liquids

Air-sensitive liquids and dry, deoxygenated solvents may be stored in any vessel capable of maintaining an internal inert atmosphere. To dispense the material by syringe techniques **(S9.2)**, it is convenient to have some kind of septum-capped inlet **(F9.36)**. Ampules[26,27] **(F9.36)** suitable for capping with rubber septa[28] are available in sizes ranging from 1 ml to 22 oz. Larger-capacity crown-cappable bottles[29] are available in sizes up to 1 gal **(F9.36B)**. The common pop bottle, which can be capped in the same manner, is a less expensive alternative. We have found crown-capped bottles with butyl rubber septa to be the method of choice for storage of compounds such as methyl borate. However, many solvent vapors decompose rubber septa and care should be exercised in storing volatile materials in this manner. In our laboratories, we routinely cap punctured

[26]They are available in 1-10 ml sizes from A. H. Thomas, Inc. Larger sizes are available by special order from Wheaton Glass, Inc.
[27]Another type of septum-capped ampule, septa, and capping equipment is available from Pierce Chemical Company in sizes from 1 to 125 ml.
[28]Rubber septa are available from Aldrich Chemical Company.
[29]Crown-cappable bottles are available in various sizes from Texas Alkyls, a subsidiary of Stauffer Chemical Company, and from Lab Glass, Inc.

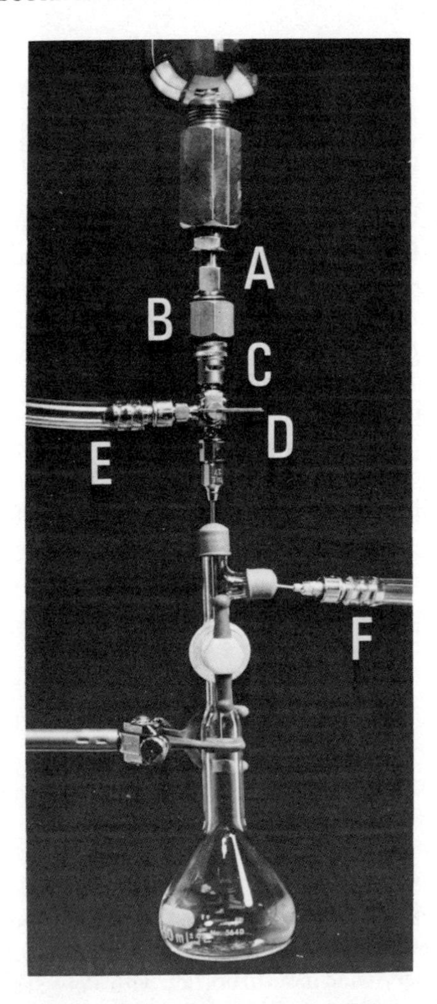

Figure 9.37. Tapping a cylinder containing an air-sensitive liquid: (A) tank valve; (B) adapter (Texas Alkyls No. 113); (C) standard thread to male Luer-Lok fitting (B-D No. 3066); (D) 3-way syringe valve (B-D No. 3156); (E) hose-barb connector to nitrogen inlet (B-D No. 9067); (F) hose-barb connector to bubbler.

septa with an inverted septum. This double-capping provides an additional seal and prolongs the life of the primary septum **(F9.40B)**.

For long-term storage of sensitive materials, vessels with a Teflon stopcock-controlled septum-capped inlet are better alternatives. A suitable round-bottom flask with adapter may be used **(F9.36C)**. However, a less expensive alternative is a large ampule **(F9.36D)**.[30]

[30]Available by special order from Ace Glass, Inc., or Wheaton Glass, Inc.

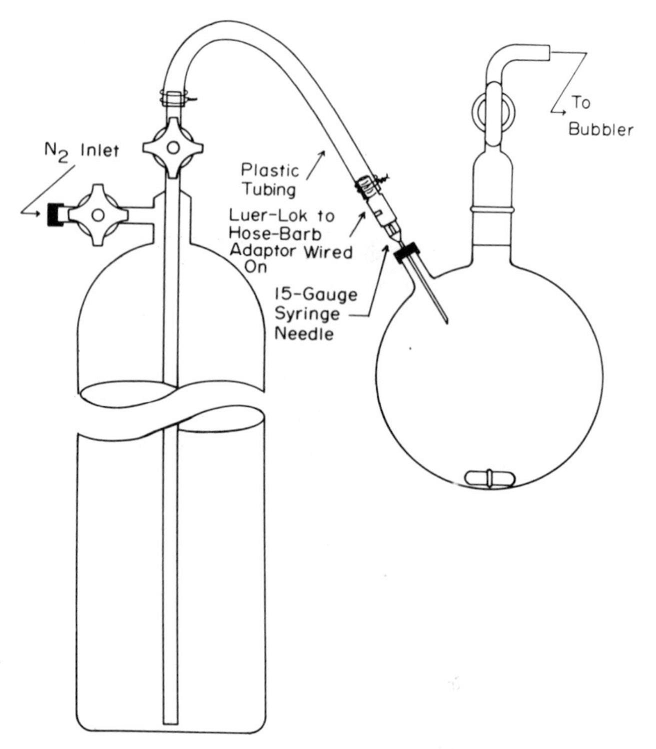

Figure 9.38. Dispensing of an organometallic from a cylinder containing an eductor tube.

Storage of Solids

Air-sensitive solids may be stored in a flask with a Teflon stopcock-controlled septum inlet **(F9.36C)**. Alternatively, crown-cappable bottles may be suitable. For short-term storage, wide-mouth bottles with Poly-Seal caps are sometimes acceptable if the outside is sealed with a pressure-sensitive tape.[31]

Commercial Packaging of Air-Sensitive Compounds

Commercially available organometallic reagents are marketed either as neat material or in solution. Commercial packages require special techniques for dispensing the contents. It is imperative that the user be familiar with both the chemistry of the material and the handling techniques required before attempting to dispense it. Most commercial suppliers will provide data sheets and handling instructions for their products.[32]

[31] 3M, Scotch Brand, No. 1471, 3VFB.

[32] See the Appendix for a partial list of suppliers and their addresses.

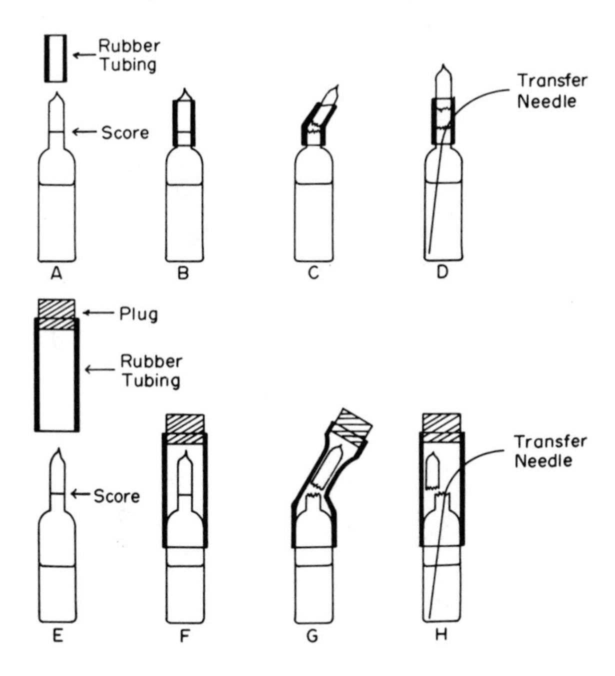

Figure 9.39. Two methods for opening an ampule under an inert atmosphere.

Neat, liquid organometallics come in two types of cylinders. One is an ordinary Lecture Bottle **(F9.37)**. The cylinder is tapped by positioning it upright and placing a septum over the outlet. The valve (A) is opened and the cylinder pressurized to 5 psi with inert gas. The valve is closed and the cylinder fitted with an adapter (B) (Texas Alkyls, No 113), a standard thread to male Luer-Lok fitting (C) (Becton-Dickinson, No 3066), a three-way syringe stopcock (D) (Becton-Dickinson, No. 3156), and a needle. The assembly is inverted and clamped to a rack. The needle is pushed through the spetum of the receiver flask and nitrogen is allowed to flow through the stopcock to flush out the needle and flask. The stopcock is opened to the cylinder and the main valve (A) *slowly* opened. After the liquid is dispensed, the cylinder valve is closed and the stopcock turned to allow nitrogen to flush out the liquid remaining in the needle.

The other type of cylinder has a eductor tube leading to the bottom **(F9.38)**. Dispensing material involves connecting a needle or some type of tubing to the center valve.[33] The cylinder is pressurized to 5 psi with inert gas through the side valve. The reagent is dispensed by allowing the pressure to force the liquid through the center valve to an appropriate receiver **(F9.38)**. When the liquid has

[33]Tygon tubing is satisfactory for trialkylboranes, but *not* for most aluminum reagents. Refer to the manufacturers' data sheets for the correct method of handling.

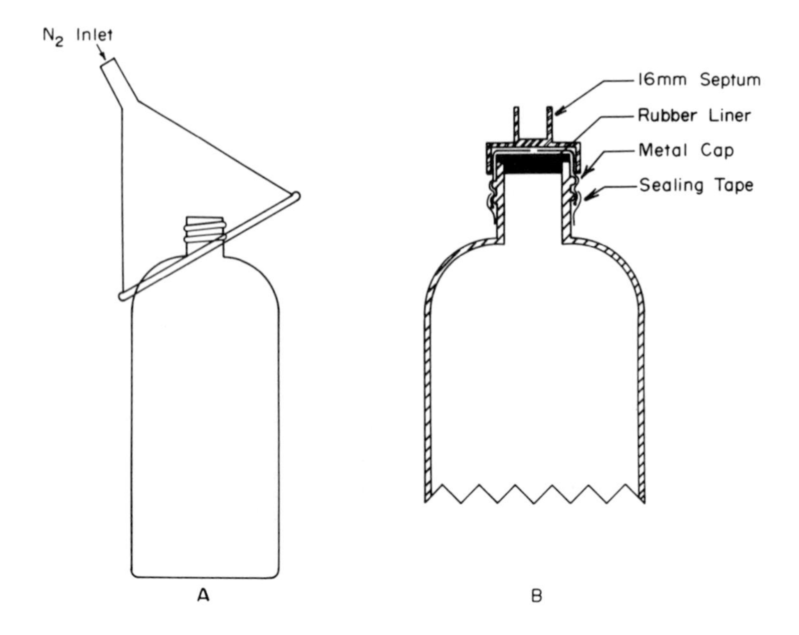

Figure 9.40. Capping a bottle under an inert atmosphere.

been dispensed, the cylinder is inverted and the excess pressure flushes the remaining liquid from the tubing and valve.[34] The protective caps, which come on the cylinder valves, should be replaced to prevent a fire in case of leakage through the valves.

Sealed ampules of organometallic material present a special problem. They are best opened in an inert-atmosphere enclosure. The ampule stem is scored with a file and broken off. A septum is placed over the opening, and the material transferred to a permanent storage vessel or the reaction flask. If no glove bag is available, the ampule may be opened in either of the following ways. The stem is scored and a tight-fitting piece of rubber tubing placed over the stem **(F9.39A-D)**. The stem is carefully broken. The rubber tubing is pulled up slightly, leaving a gap between the two pieces of glass. The material may be transferred by inserting a syringe needle through the rubber tubing. Alternatively, for prescored ampules, a larger piece of tubing may be slipped over the whole ampule **(F9.39E-H)** and capped with an appropriate stopper. A nitrogen needle and vent needle are inserted and the head space flushed with nitrogen. The vent

[34]*CAUTION:* Enough space must remain in the receiver to allow the material in the transfer and eductor tubes to be flushed out.

needle is removed and the stem carefully broken. The reagent may then be removed by syringe.

Solutions of organometallics often come in screwtop bottles with Poly-Seal caps. These caps must be replaced with septum closures to permit access. This can be carried out in a glove bag. It may also be done on the bench using an inverted funnel to provide a blanket of inert atmosphere (**F9.40**). The cap is quickly removed (*CAUTION*! the contents may be under excessive pressure) and the septum liner placed on the top. A metal cap with a hole is screwed down to hold the liner in place. The cap is sealed to the bottle with pressure-sensitive tape.[31] A 16-mm sleeve-type stopple is inverted, pulled down over the cap, and wired on. In use, a syringe needle is pushed through both septa to gain access to the reagent.

9.10. ANALYSIS OF BORANES, ORGANOBORANES, AND ORGANOMETALLICS[35]

NMR Spectra

NMR spectra of air-sensitive materials can be obtained easily. Small septums are available for sealing the open end of 5-mm NMR tubes. The tube may be flushed with nitrogen, using a long needle. Samples are then transferred into the tubes, using ordinary syringe techniques. Since NMR analysis is very susceptible to small amounts of particulate matter, a filter assembly, such as that shown in **F9.41**, may be used. This assembly consists of a hypodermic syringe, a three-way valve, and a Swinney filter (Becton-Dickinson, No. 3061) equipped with a 10-mm glass or polyethylene frit. The sample is forced through the filter into the syringe by nitrogen pressure. The stopcock is turned and a measured amount of the liquid transferred into the NMR tube. This technique is limited to filtration of small amounts of solid due to the low capacity of this filter.

^{11}B NMR spectra are commonly obtained on internally locked spectrometers, which require a locking liquid. This liquid cannot always be mixed with reactive samples. **F9.42** shows one method for handling this problem. The locking liquid or the reference material (BF_3:Et_2O) is placed in the 10-mm tube, and the sample is placed in the 5-mm septum-capped tube. The spacers are made from vortex plugs (Wilmad Glass, Inc.) which have been bored out in a lathe using a No. 8 drill and tapped with 1/4-20 threads. The machining must be done carefully so that the tube will spin properly. A 1/4-20 threaded aluminum or plastic rod is

[35] For an excellent account of analysis of organometallics see T. R. Crompton, *Chemical Analysis of Organometallic Compounds, Vol. I, Elements of Groups I-III*, Academic Press, New York, 1973.

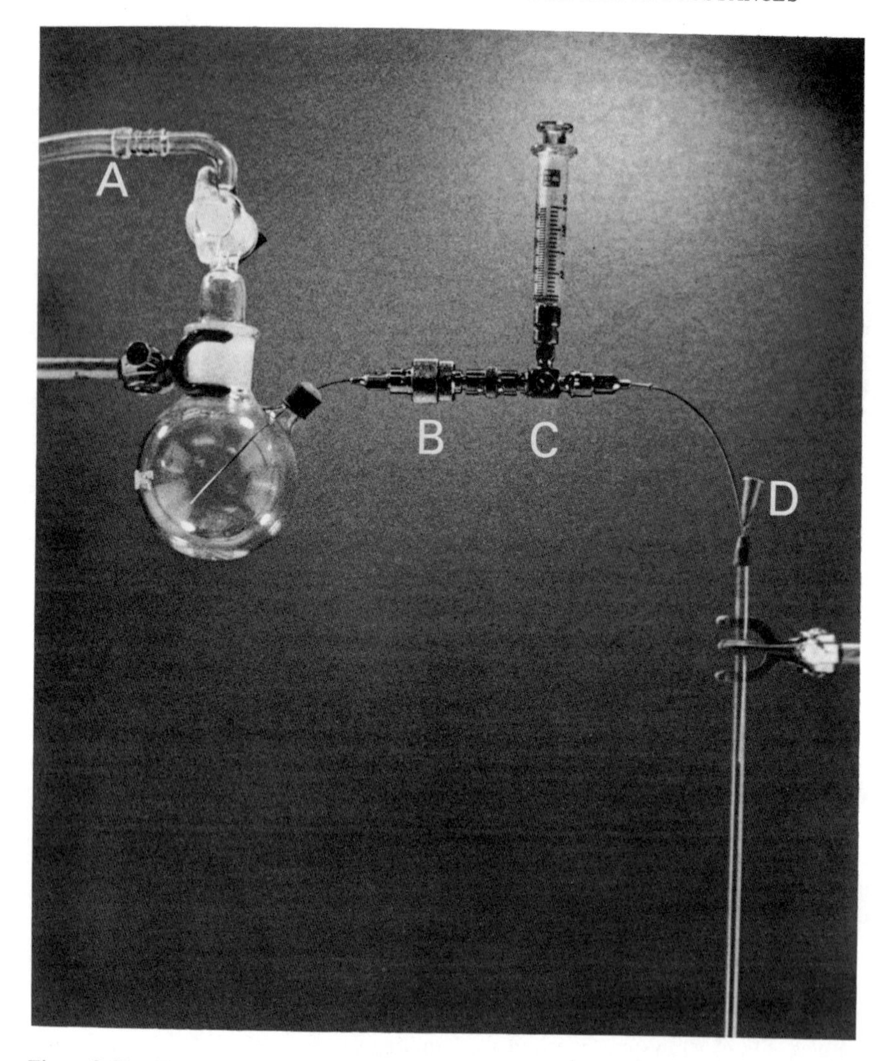

Figure 9.41. Filtration of an NMR sample: (A) nitrogen inlet; (B) Swinney filter (B-D No. 3061); (C) 3-way valve (B-D No. 3158 or 3156, and adapter No. 3114; (D) vent needle.

used to position the spacers in the 10-mm tube. Since ^{11}B NMR shifts are large, errors due to bulk susceptibility effects in this assembly are negligible.

Infrared Spectra

Problems arising in taking infrared spectra of air-sensitive liquids can be minimized by using the two-syringe technique, shown in **F9.43**. The IR cell and the

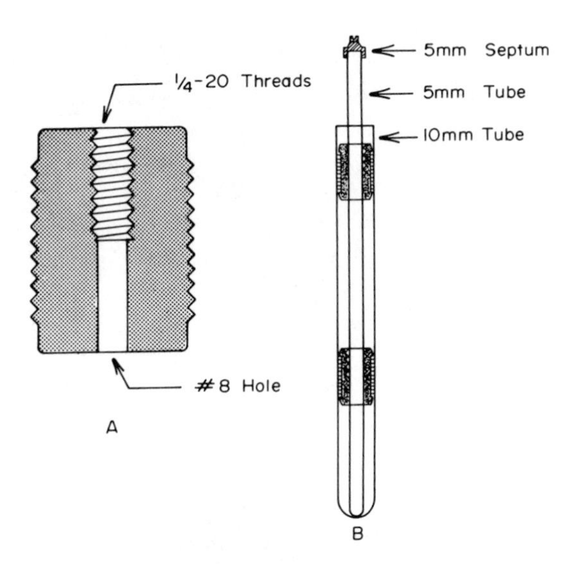

Figure 9.42. NMR tube holder. A modified vortex plug (Wilmad No. WG-805A) serves as a concentric spacer.

reservoir syringe are flushed with nitrogen and the reservoir syringe left filled with nitrogen. Using a syringe with a stopcock, the sample is withdrawn in the usual manner. The stopcock is closed and the needle removed. The sample syringe/stopcock assembly is quickly attached to the cell, while the reservoir syringe plunger is depressed, giving a final purge to the cell. The stopcock is opened, and some of the sample is pushed through the cell into the reservoir syringe, rinsing the cell with the sample. The stopcock is closed and the spectrum recorded. Then the stopcock is opened and the reservoir syringe plunger depressed, forcing sample back into the sample syringe. Normally, sample losses using this procedure are quite low. In addition to the obvious advantages, this technique provides a means of pressure release, if the sample expands or evolves a gas, during the spectral analysis. It is quite common to have a small amount of hydrogen evolved from solutions of active hydrides. In a sealed cell, this pressure increase can cause the cell gaskets to leak.

Infrared spectra of air-sensitive solids are best obtained in solution. Where this is undesirable or impossible, the solid can be mulled with Nujol or Florolube in an inert-atmosphere enclosure. The mull is applied to the salt plates in the usual manner, but before leaving the inert atmosphere, the salt plates are sealed on the edges with a sealing tape.[31] The spectrum is recorded in the usual manner.

Standaridization of Metal Hydride Solution

Borohydrides and other metal hydrides may be hydrolyzed rapidly and quanti-

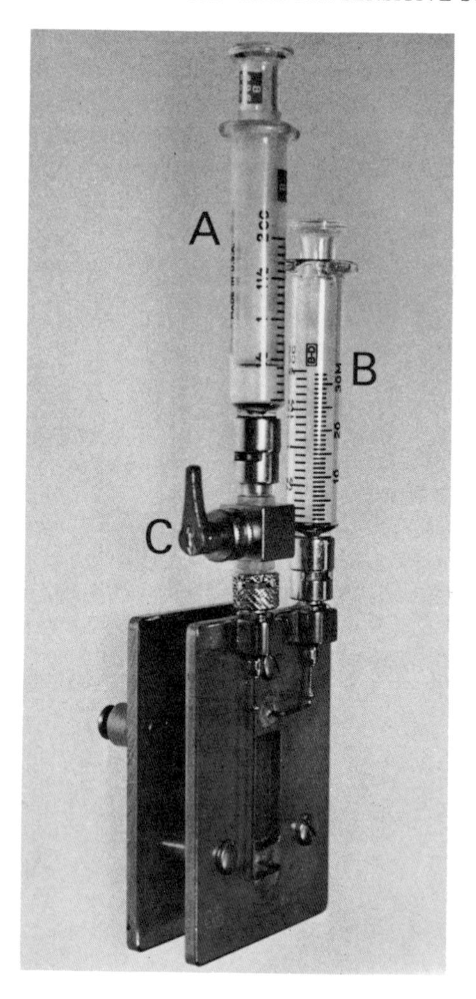

Figure 9.43. Sample cell for IR spectra of air-sensitive solutions: (A) sample syringe; (B) reservoir syringe; (C) syringe valve (Hamilton No. 1LF1, 86570).

tatively, liberating 1 mole of hydrogen per hydride. This characteristic property provides a convenient method for their analysis. The hydrogen evolved may be measured in a gas buret such as that shown in **F9.44**. A 250-ml buret is normally used, but for dilute solutions, a smaller buret is more practical.

The buret is filled with distilled water containing a small amount of cupric sulfate which makes the liquid level easier to read. The hydrolysis flask is filled with about 100 ml of a hydrolyzing solution (**T9.2**). The condenser and Dry-Ice trap prevent solvent vapors from contaminating the water, thereby changing its vapor pressure.

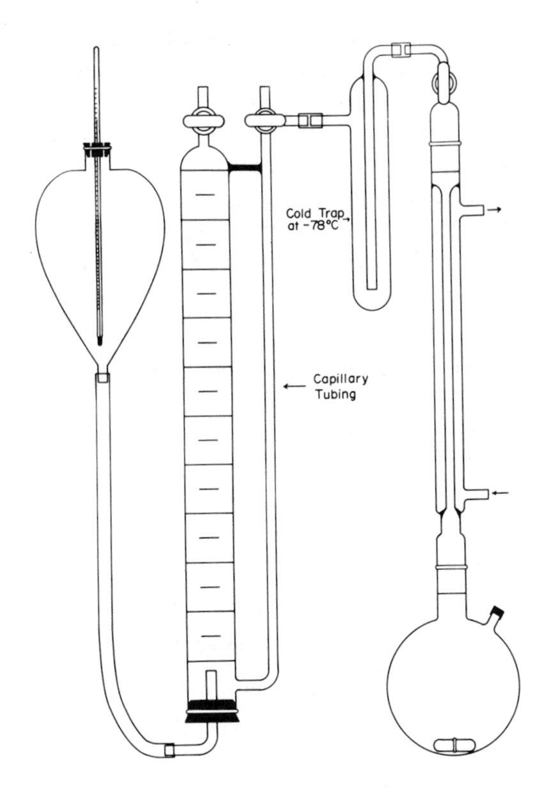

Figure 9.44. Analytical gas buret.

The hydrolysis of borane with water and the lower alcohols proceeds in stages. Two hydrides react rapidly, but the third is so slowly hydrolyzed that the intermediate compound, $(RO)_2B$-H, can be isolated.[36] However, in the presence of a polyglycol, such as glycerol or mannitol, all three hydrides are readily displaced, presumably because of a chelation effect. Unlike borane itself, partially alkylated boron hydrides react rapidly with water and the lower alcohols with complete displacement of all their B-H bonds. 9-BBN is somewhat of an exception, since it precipitates in water-THF and is hydrolyzed very slowly. However, it is easily methanolyzed at 25°, but at a slower rate than most other dialkylboranes.

Commercial BH_3:THF contains about 5 mole % dissolved sodium borohydride (20 mole % hydride) as a stabilizer. As the hydrogens on commercial borane are replaced with alkyl groups, the relative amount of hydride in solution due to the

[36] A. N. Nesmeyanov and R. A. Sokolik, *Methods of Elemento-Organic Chemistry,* Vol. 1, North-Holland Publishing Co., Amsterdam, 1967.

sodium borohydride increases from ~6.6% for BH_3 to ~10% for RBH_2 and ~20% for R_2BH. If the sodium borohydride is completely hydrolyzed in the analysis, a sizable error is introduced. Fortunately, sodium borohydride is only partially hydrolyzed in THF-water-glycerol, so that only a tailing endpoint is observed. This can be tolerated for BH_3:THF, since the hydrolysis is so fast that a reliable determination can be made before much sodium borohydride reacts. Since sodium borohydride does not react rapidly in ethanol-THF, mono- and dialkylboranes can be hydrolyzed in this medium to give well-defined end points.

Table 9.2. Conditions for Hydrolyzing Metal Hydride Solutions

Hydride	Hydrolysis Solution (1:1)
BH_3:THF	Glycerol-water
BH_3:$S(CH_3)_2$	Glycerol-water-methanol
BH_2Cl:$O(C_2H_5)_2$	Glycerol-water
$BHCl_2$:$O(C_2H_5)_2$	Glycerol-water
9-BBN	Methanol-THF[a]
$Li[R_3BH]$	Glycerol-water-THF
$LiAlH_4$	2*N* H_2SO_4-THF
$LiAl(OCH_3)_3H$	2*N* H_2SO_4-THF
AlH_3	2*N* H_2SO_4-THF
$LiBH_4$	Glycerol-water
$NaBH_4$	2*N* HCl
RBH_2, R_2BH	ethanol-THF

[a]The hydrolysis is slow at 25° but gives better results at 35°.

An accurately measured aliquot of the hydride solution is injected into the hydrolysis mixture. The gas evolved should be sufficient to fill at least 50% of the buret. The concentration is then determined from the following equation:

$$\text{hydride molarity} = \frac{(P_1 - P_{H_2O})(273)(V_{H_2} - V_B)}{760\,(T)(22.4)(V_A)}$$

where P_1 = atmospheric pressure (mmHg)
P_{H_2O} = vapor pressure of water at T
V_{H_2} = volume of hydrogen measured

V_A = volume of aliquot injected
V_B = volume displaced by aliquot
T = water temperature in buret (°K)

When dilute hydride solutions are analyzed, the displaced volume, V_B, may not be equal to the amount of hydride solution added, V_A, due to mixing effects. Therefore, a separate injection of the solvent must be made to determine V_B.

Procedure

The buret stopcock is opened and the three-way stopcock turned to connect the buret to the hydrolysis flask. The system is saturated with hydrogen by injecting a few milliliters of the hydride solution. The three-way stopcock is turned so that the system is open to the air and the water level adjusted with the leveling bulb to give a near-zero reading. The height of the water level in the capillary tube above the level in the buret due to capillary action is noted. The buret stopcock is closed and the three-way stopcock turned to connect the buret and hydrolysis flask. An accurately measured aliquot of the hydride solution is injected into the hydrolysis solution **(N1)**. The hydrogen evolution may be monitored by following the level of the water in the capillary tube **(N2)**. When no more gas is evolved, the level of the water in the leveling bulb, buret, and capillary tube (minus the height due to capillary action) must be equalized. If necessary, the leveling bulb is lowered to just below the bottom of the buret to force a small amount of gas into the buret. The final reading is then taken along with the water temperature and atmospheric pressure and the concentration calculated. The hydrolysis is repeated with several aliquots **(N3)**.

Notes

1. The syringes must be thoroughly dried. It is often advisable to lubricate the barrel with an inert oil to prevent the plunger from sticking. Any moisture on the syringe barrel, even fingerprints, may cause trouble.
2. A ⅛-in. section of rubber tubing cut to fit around the capillary tube provides a handy reminder of the liquid level.
3. With good technique, the precision and accuracy is ±2%. Several determinations are generally made until reproducible values are obtained. The first values are often erratic when fresh hydrolysis medium is used.

Titration of Boric Acid

Boric acid is a very weak acid and gives a poor endpoint when titrated with base. However, the addition of a chelating agent, such as mannitol, causes the boric acid to behave as a stronger acid. Consequently, it may be titrated with

$$B(OH)_3 + \begin{matrix} HO- \\ | \\ HO- \end{matrix} \rightarrow \left[\begin{matrix} HO & & O- \\ & B & \\ HO & & O- \end{matrix} \right]^- + H^+ + H_2O$$

standard sodium hydroxide. A strong acid may be titrated in the presence of boric acid using methyl orange as an indicator. Then about 2 g of mannitol is added for each 5 mmoles of boric acid and the solution titrated to the phenolphthalein endpoint.

Some boronic acids, $RB(OH)_2$, may also be titrated in the presence of mannitol using sodium hydroxide and phenolphthalein.[36] However, borinic acids, R_2BOH, usually give poor endpoints. The endpoint is sharper if the solution is briefly heated before titration.

Analysis of Organoboranes by VPC[37]

Lower molecular weight organoboranes are relatively volatile and may be analyzed by gas chromatography. Their retention time on a nonpolar column is approximately the same as that of the corresponding hydrocarbon. However, because of their reactivity, certain precautions should be observed.

Inert liquid and stationary phases should be used for column packing. SE-30, Apiezon-L, or similar silicon or hydrocarbon-based liquid phase on Chromosorb W, AW, DMCS, works well. Normally a 6-ft column is used. Forlabile organoboranes, a shorter column will give better results if the separation is not difficult. Protic sites must be masked with trimethylsilyl groups by injecting four to five 50-μl quantities of Silyl-8 (Pierce Chemical Co.) at 150-250°. Normally a few 50-μl injections of the organoborane or triethylborane are required to remove oxygen and to obtain reproducible results.

Other compounds should not in general be analyzed on columns used for organoboranes. Boron containing residues may seriously interfere with certain compounds, especially alcohols and amines. A single VPC or one channel of a dual column instrument should be reserved for organoborane analysis. Air and species containing boron-hydride bonds should be excluded from the column during use.

The injection block temperature should be cool enough to prevent thermal decomposition of the organoborane. For organoboranes, such as the tributylboranes, normal injection port temperatures ($\leqslant$250°C) may be used. However for thermally sensitive organoboranes, the injection port temperature must be reduced.

Organoboranes may be quantitated by VPC once a relative thermal response factor has been determined. An accurately weighed sample of the organoborane

[37] For an excellent review of gas chromatography separations and techniques, see H. M. McNair and E. J. Bonelli, *Basic Gas Chromatography*, Varian Aerograph, 1968.

and internal standard (usually a normal hydrocarbon) in a suitable solvent is prepared under nitrogen. The response factor should be determined several times to ensure reproducibility. The following equation may be used for the response factor (RF):

$$\text{RF} = \frac{\text{mmole product}}{\text{mmole standard}} \times \frac{\text{area standard}}{\text{area product}}$$

When exploratory reactions are being studied, it is often informative to follow the loss of an organoborane by VPC.

The purity of isomeric organoboranes may be checked by VPC. For example, hydroboration of 1-butene places 94% of the boron in the 1-position and 6% in the 2-position. The organoborane consists of a statistical mixture of 83% tri-*n*-butylborane, 16% di-*n*-butyl-*sec*-butylborane, 1% *n*-butyldi-*sec*-butylborane and a trace of tri-*sec*-butylborane. These isomers are easily separated on a ¼-in × 6-ft SE-30 column.

Analysis of Organoboranes by Oxidation

When a large number of exploratory or repetitive reactions are to be carried out using a single organoborane, it is often convenient to prepare a stock solution of the organoborane. The olefin may be hydroborated in THF and then standardized by oxidation with alkaline-hydrogen peroxide. This process also serves as a check on the isomeric purity (tri-*sec*-butylborane **P2.7**) and structure of the organoborane. For more difficult to oxidize organoboranes, see **P6.10**.

$$R_3B + NaOH + 3\,H_2O_2 \longrightarrow 3\,ROH + [B(OH)_4]Na$$

Procedure

A dry 50-ml flask fitted with a reflux condenser, septum inlet, and magnetic stirring bar is flushed with nitrogen. The organoborane (10.0 ml of an approximately 0.5*M* solution) is placed in the flask by syringe. The flask is immersed in an ice bath, and 2.0 ml of 3*M* sodium hydroxide is added. Hydrogen peroxide, 2.0 ml of a 30% solution, is added at such a rate to keep the temperature below 50° (**N1**). After the addition, the reaction is stirred an additional 10 minutes at 40-50° (warm-water bath), then cooled to room temperature. The aqueous phase is saturated with potassium carbonate (**N2**). An appropriate internal standard is added and the solution analyzed for alcohols by VPC (**N3**).

Notes

1. *CAUTION!* The addition of hydrogen peroxide often causes an extremely exothermic reaction. The peroxide should be added at a slow rate to keep

the solution below the reflux point of THF. If a vigorous reaction does not occur with addition of the first few drops of peroxide, the ice bath should be replaced with 40-50° water bath. The oxidation normally requires 1.67 ml of 3*M* sodium hydroxide and 1.67 ml of 30% hydrogen peroxide for 5 mmoles of trialkylborane. An excess is used here in case the solution is greater than 0.5*M*.

2. If excess hydrogen peroxide is present, some foaming will occur. The potassium carbonate must be added slowly.
3. The choice of VPC column and internal standard depends on the mixture being analyzed. The yield of alcohol is determined by integrating the area of the alcohol and internal standard and comparing to the areas for an authentic mixture of known amounts.[37]

Organolithiums and Grignard Reagents

Standardized organolithium and Grignard reagents are available from a variety of commercial sources. However, the concentration may not be accurate and may change with time. Therefore, the solutions should be standardized periodically. The method of Watson and Eastham[38] which directly titrates the metal-carbon bond with *sec*-butyl alcohol is an excellent procedure. A charge-transfer complex between the organometallic and 1,10-phenanthroline or 2,2′-biquinoline is used as an indicator. 1,10-Phenanthroline is a good indicator for lithium reagents, but certain Grignards, such as allylmagnesium chloride in THF, give a better color with 2,2′-biquinoline. The total base, organometallic plus alkoxide, may be determined by quenching an aliquot with 5*M* water in THF and then titrating with standard 0.2*M* HCl to a bromothylmol blue endpoint.

The titration must be carried out entirely under nitrogen. A typical setup is shown in **F9.45**.

Procedure

An oven-dried 50-ml Erlenmeyer flask capped with a wired-on septum and containing a magnetic stirring bar is flushed with nitrogen. To ensure dryness, the flask is flame-dried during the nitrogen flushing. About 20 ml of dry benzene **(N1)** and 0.2-0.5 ml of indicator solution **(N2)** are placed in the flask. Too much indicator may give a poor endpoint. The dry buret is flushed with nitrogen and filled with 1.0*M* *sec*-butyl alcohol in xylene **(N3)**, then inserted through the prepunctured serum stopple. An aliquot of organometallic (3.00 ml for a 2*M* solution) is added to the benzene and the solution titrated until the color disappears **(N4)**.

[38] S. C. Watson and J. F. Eastham, *J. Organomet. Chem.*, **9**, 165 (1967).

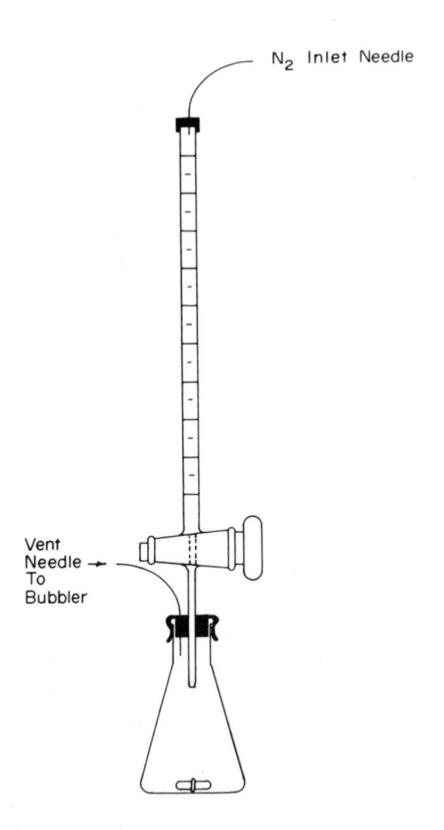

Figure 9.45. Titration of air-sensitive samples.

Notes

1. Spectro-grade benzene stored under nitrogen over 5 A Molecular Sieves is used.
2. 100 mg of indicator is placed in 50 ml of dry benzene under nitrogen.
3. *sec*-Butyl alcohol is distilled from calcium hydride under nitrogen. Xylene is distilled from LAH under reduced pressure or from sodium. Reagent-grade xylene stored over Molecular Sieves may also be used. A standard solution may be prepared by adding an accurately weighed sample of alcohol to a volumetric flask and diluting to volume under nitrogen.
4. A reproducibility of ±1-2% can be expected with good technique.

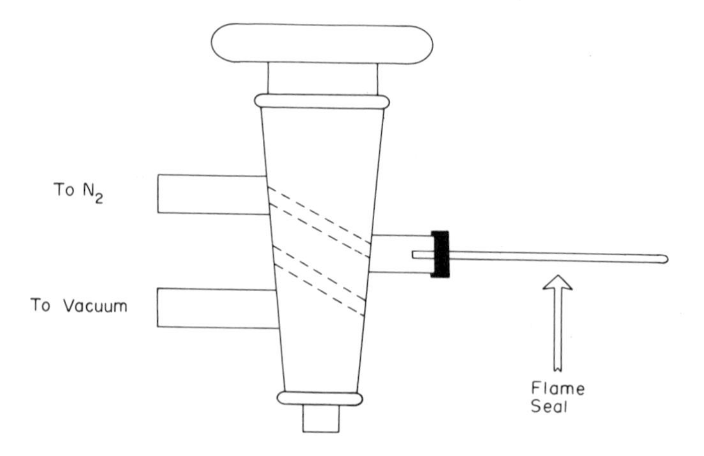

Figure 9.46. Preparing a sealed, evacuated melting-point tube.

Melting Points of Air-Sensitive Materials

Melting points of air-sensitive materials are commonly obtained in sealed-evacuated capillaries.[39] The open end of a melting point capillary is inserted through the prepunctured septum on the stopcock (**F9.46**). The tube is flame dried while being purged with nitrogen by the alternate evacuation-refilling method. With the stopcock closed, the assembly is transferred to an inert atmosphere enclosure where the capillary is removed, filled with the sample and reinserted into the septum. The apparatus is removed from the enclosure and connected to a vacuum source. After the capillary has been evacuated, it is sealed above the sample. The seal should be made close enough to the sample so that the entire enclosed portion will be within the heated zone of the melting point apparatus. This prevents sublimation of the sample to a cooler portion of the tube during heating.

Melting points of boronic and borinic acids are particularly difficult to obtain since these compounds are easily thermally dehydrated. Even partial dehydration can lead to erroneous melting points.[36] A successive approximation technique has been developed to overcome this problem by minimizing the contact time of the sample with the hot apparatus.

First, a sample is heated rapidly (10-15°/minute) until it melts. The heating unit is then cooled about 10° below this point where a second sample is heated

[39]Ref. 20, p. 242.

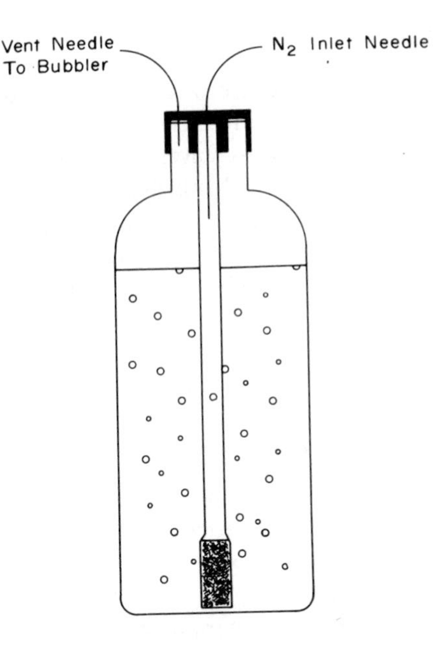

Figure 9.47. Degassing a solvent with a gas-dispersion tube.

more slowly to melting. The process is continued until a reproducible melting point is obtained. Each time the starting temperature is raised and the heating rate is lowered. Caution should be observed when using melting points to characterize boronic and borinic acids, since there is usually no assurance that reported melting points were obtained in this manner.

9.11. PURIFICATION OF SOLVENTS AND REAGENTS

Many organometallics and hydrides are susceptible to oxygen and water. Consequently, it is essential that the solvents be free of these reactive impurities. For example, a 1*M* BH_3:THF solution made with THF containing 2% water will lose one equivalent of hydride. Furthermore, the residual borane (possibly a $HOBH_2$ species) may react far differently than borane itself, providing poor yields of the desired organoborane and very poor yields of the desired product in subsequent reactions. In utilizing more dilute solutions, even much smaller amounts of impurities in the solvent can be highly detrimental. Therefore, the purification, storage, and handling of solvents for reactions involving organometallics and hydrides should be carried out with great care.

Reagent-Grade Solvents

Anhydrous reagent-grade solvents are often adequately pure for direct use. To ensure that these materials are sufficiently dry, they may be stored over activated Molecular Sieves.[40,41] Most solvents can be partially deoxygenated by bubbling nitrogen through the liquid utilizing a needle, or, more efficiently, a gas-dispersing tube.[42] **F9.47** shows a convenient method for deoxygenation using a gas-dispersion tube which is held in place with the rubber septum that caps the bottle.[43] The dispersion tube is left in the bottle and the solvent is withdrawn as needed by inserting a syringe needle through the septum into the dispersion tube. The dispersion tube thus can serve as a filter to remove small particles of desiccant.

Drying and Deoxygenating Solvents

Most solvents can be purified by direct distillation from a carefully selected drying agent (**T9.3**) under an inert atmosphere (**S9.7**).[44] For all reactions in this book, solvents purified in the above manner are satisfactory. However, other reactions may be affected by extremely small amounts of oxygen. Distillation from a drying agent under an inert atmosphere is not sufficient. Residual oxygen must be removed by physical or chemical means.

Oxygen may be removed physically by the freeze-pump-thaw method.[45] Unfortunately, this technique is both time consuming and cumbersome when large amounts of solvent are involved. Therefore, if possible, a chemical oxygen scavenger is preferred. Distillation of the solvent from a benzophenone ketyl is

[40] C. K. Hersh, *Molecular Sieves,* Reinhold Publishing Corp., New York, 1961; R. F. Gould (ed.), *Molecular Sieve Zeolites,* I and II, Advances in Chemistry Series, No. 101, 102, American Chemical Society, Washington, D.C., 1971.

[41] New Molecular Sieves can be activated by heating at 320° for 3 hours and cooling in an evacuated desiccator. The desiccator should be returned to atmospheric pressure with nitrogen to prevent entrainment of air in the sieves. Used Molecular Sieves can be regenerated by first removing any residual solvent (by heating the sieves in a water aspirator-pumped flask), followed by heating in a well-ventilated oven for 12 hours at 320°. Ether solvents stored over Molecular Sieves should be maintained under nitrogen since anhydrous ethers form peroxides readily.

[42] K. Koyma and C. Michelson, *Anal. Chem.,* **29**, 1115 (1957).

[43] Personal communication, M. E. Swanson and R. E. Jensen, Purdue University.

[44] For a more complete discussion of solvent purification, see: (a) Ref. 2; (b) J. A. Riddick and W. B. Banger, *Techniques of Chemistry,* Vol. 2; *Organic Solvents: Physical Properties and Methods of Purification,* 3rd ed., Interscience, New York, 1971; (c) A. J. Gordon and R. A. Ford, *The Chemist's Companion,* Wiley-Interscience, New York, 1972; (d) J. Burlitch, Technical Bulletin No. 570, Ace Glass Inc., Vineland, N.J.

[45] A. Nolle and P. Mahendroo, *J. Chem. Phys.,* **33**, 863 (1960).

Table 9.3. Reagent and Solvent Purification

Solvent or Reagent	Chemical Pretreatment	Drying Agent
Saturated Hydrocarbons	Conc. H_2SO_4[a]	CaH_2, LAH,[d] or Ketyl[e]
Olefins[b,c]	–	LAH[d] or CaH_2
Aromatics	Conc. H_2SO_4[f]	CaH_2, LAH,[d] or Ketyl[e]
Et_2O	Sodium or 3 A Molecular Sieves	CaH_2, LAH,[d] or Ketyl[e]
THF	5 A Molecular Sieves or CaH_2	LAH,[d] or Ketyl[e]
Glymes: DME, diglyme, etc.[g]	CaH_2, or sodium	LAH,[d] or Ketyl[e]
$CHCl_3$, CH_2Cl_2, CCl_4	Conc. H_2SO_4[a]	P_2O_5[h]
MeOH	3 A Molecular Sieves	Magnesium
Pyridine	KOH	BaO, CaH_2
DMF	4 A Molecular Sieves	BaO
DMSO	4 A Molecular Sieves	CaH_2, BaO

[a] Solvent stirred over the acid for several days.
[b] Olefins form peroxides on standing for long periods in contact with air. If the olefin requires a large amount of LAH for drying and destroying the peroxide, the dried material should be filtered before distillation to lessen the chance of an exothermic polymerization catalyzed by aluminum salts.
[c] Vacuum distillations of many olefins from LAH are often plagued by excessive foaming. Proper precautions (splash guards or long Vigreux columns) should be taken.
[d] *CAUTION:* Solutions of LAH in oxygen-containing solvents may decompose explosively at elevated temperatures (~160°). Therefore the distillation should never be carried out to dryness, and materials boiling above 100° should be distilled under reduced pressure.
[e] Sodium or potassium benzophenone ketyl.
[f] Optional for thiophene-free grades.
[g] Commercial glymes are usually wet and often contain relatively large amounts of peroxides. These materials should be predried by stirring with CaH_2 overnight or heating to reflux with sodium.
[h] Polyhalogenated materials should never be contacted with sodium or LAH because of the danger of an explosion.

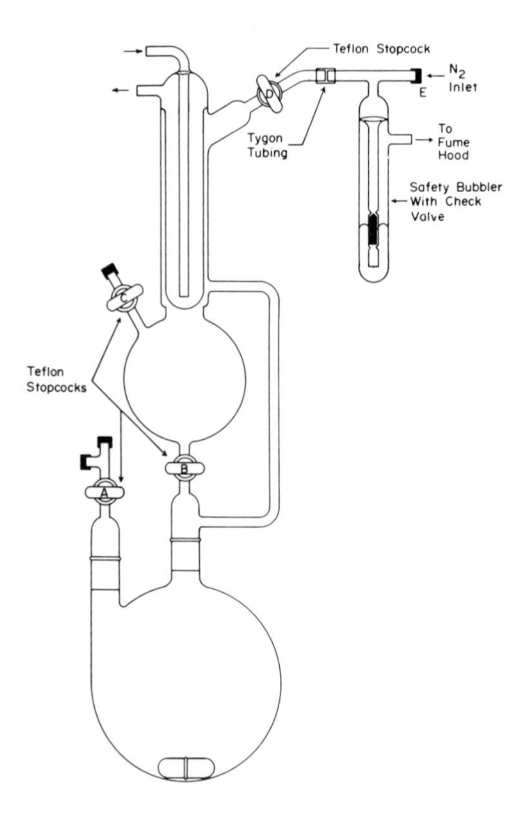

Figure 9.48. Solvent still.

an effective method for drying and deoxygenating many solvents.[44c] The solvent is refluxed with benzophenone and potassium or sodium until the characteristic purple color of the ketyl persists. The solvent, which is water- and oxygen-free, is then distilled as usual. An alternative procedure for oxygen and peroxide removal involves passing the solvent through a column of reduced BTS catalyst.[46] After this deoxygenation step, the solvent must be distilled from an appropriate drying agent.

Solvent Still

It is convenient to set up a permanent still for solvents which are routinely used in large amounts. A practical design is shown in **F9.48**. The receiver-condenser

[46]BTS catalyst and descriptive literature is available from Badische Aniline und Soda-Fabrik and Ace Glass Co.

assembly is entirely glass and Teflon. The glassware is oven-dried, assembled hot, and allowed to cool with nitrogen flowing through a long needle in stopcock A and out through stopcocks B and D. When the apparatus is thoroughly flushed, the nitrogen needle is removed from A and inserted through septum E on the bubbler to provide the usual static pressure of nitrogen. Predried solvent is added to the pot through A with a double-ended needle.[47] The drying agent is introduced through the side neck by removing temporarily the connecting tube A.[48] After the drying agent is added, the solvent is heated to reflux with stopcocks B and D open. This reflux is maintained for a few minutes to remove any traces of moisture from the receiver and condenser surfaces. Stopcock B is then closed, and the receiver allowed to fill. The distillate is transferred with a double-ended needle through stopcock C into a storage container. Additional solvent may be added to the pot through stopcock A. After any addition of solvent, the pot material should be retested for active hydride. Before the still is disassembled, a small quantity of hexane is added and any excess hydride destroyed by the slow addition of ethyl acetate.[49]

Purification of Ethyl Ether

Anhydrous, reagent-grade ethyl ether is suitable for most purposes (**N1**). Technical-grade ether may be purified by distillation in an inert-atmosphere still (**S9.7**). Approximately 1 ℓ of technical-grade ethyl ether is stored over activated Molecular Sieves (3 A) for 1-2 days (**N2**). The predried ether is transferred by double-ended needle (**S9.2**) into an inert-atmosphere still (**F9.48, S9.7**). Lithium aluminum hydride (**N3**) is added under a blanket of nitrogen in small amounts using a *plastic* spatula. A sufficient amount of LAH has been added when a 1-ml aliquot of the solvent shows vigorous hydrogen evolution on hydrolysis with a drop of water (**N4**). The ether is heated to reflux with a gentle flow of nitrogen through the system, then the nitrogen needle placed in the bubbler

[47] Certain solvents should be pretreated before final distillation (**T9.3**). Predrying eliminates the need for large amounts of desiccant in the final distillation.
[48] If lithium aluminum hydride is used as the drying agent, it should be added in *small* amounts using a *plastic* spatula. LAH is advantageous since it also reduces peroxides present in many solvents. One should not assume that bubbling observed on addition of solid LAH to the solvent necessarily indicates that the solvent is still wet. In certain solvents, such as glyme (DME) and diglyme, bubbling occurs on mixing solid LAH even with very dry material. Sufficient LAH has been added when a 1-ml aliquot of the solvent shows vigorous hydrogen evolution on hydrolysis with a drop of water. If the solvent requires more than 1% by weight of LAH, the solvent should be predried with other drying agents prior to final distillation from LAH (see **T9.3**). Solutions of LAH in oxygenated solvents can decompose explosively at temperatures above 160°. Therefore, the material in the pot should never be distilled to dryness, and less volatile solvents should be distilled under reduced pressure to keep the temperature below 100°.
[49] Ref. 20, p. 584.

(F9.48) and the solvent collected, bp 34-35°.

Notes

1. The ether may be stored over Molecular Sieves and flushed with nitrogen as previously described.
2. Anhydrous ether begins to form peroxides within 3 to 4 days and should *not* be allowed to stand over Molecular Sieves for long periods without protection from the atmosphere.
3. Diethyl ether may bubble upon the addition of LAH due to the heat of solvation. Therefore, the *best* way to test for active hydride is to hydrolyze an aliquot.
4. No more than 1% by weight of LAH should ever be required.

Purification of Tetrahydrofuran

Anhydrous reagent-grade THF is satisfactory for many purposes **(N1)**. Technical-grade THF may be purified in an inert-atmosphere still. Approximately 1 ℓ of technical-grade material is predried over Molecular Sieves (5 A) or calcium hydride for 1-2 days **(N2)**. The THF is transferred into an inert-atmosphere still **(F9.48, S9.7)** by double-ended needle **(S9.2)**. Lithium aluminum hydride is added under a blanket of nitrogen in small amounts using a *plastic* spatula. A sufficient amount of LAH has been added when a 1-ml aliquot of the solvent shows vigorous hydrogen evolution on hydrolysis with a drop of water **(N3)**. The THF is heated to reflux with a slow flow of nitrogen through the system. The nitrogen needle is placed in the bubbler to provide a static pressure of nitrogen and the THF collected in the normal manner, bp 65-67°.

Notes

1. Reagent-grade THF may be dried over Molecular Sieves (5 A) and flushed with nitrogen as previously described.
2. Anhydrous THF rapidly forms peroxides (1-2 days) upon exposure to oxygen. Therefore, THF should not be stored over Molecular Sieves for long periods of time.
3. Addition of LAH to THF leads to bubbling even when the THF is dry! Therefore, the *best* way to test for active hydride is to hydrolyze an aliquot. No more than 1% by weight of LAH should ever be required using this method.

Purification of Diglyme[50]

Commercial diglyme is usually somewhat wet. For purification, approximately

[50] G. Zweifel and H. C. Brown, *Organic Reactions,* Vol. XIII, John Wiley and Sons, Inc., New York, 1963, p. 1.

1 ℓ of technical-grade diglyme is dried by stirring over calcium hydride for a day (**N1**). The predried diglyme is transferred (**S9.2**) into an inert-atmosphere still (**S9.7**) set up for distillation under vacuum. Lithium aluminum hydride is added under a blanket of nitrogen in small amounts using a *plastic* spatula. A sufficient amount of LAH has been added when a 1-ml aliquot of the solvent shows vigorous hydrogen evolution on hydrolysis with a drop of water (**N2**). The diglyme is distilled in the normal fashion under vacuum, and the vacuum released with nitrogen (**S9.7**), bp 60° at 15 mm.

Notes

1. Anhydrous diglyme will form peroxides upon standing in contact with the atmosphere.
2. Anhydrous diglyme bubbles vigorously upon addition of LAH. Therefore, the *best* method of testing for active hydride is to hydrolyze an aliquot. No more than 1% by weight of LAH should ever be required using this method.

Purification of Olefins

Olefinic reagents must be dry and peroxide-free for successful use in the hydroboration reaction. Olefinic hydrocarbons may be purified by distillation under nitrogen from a small amount of active LAH. Lithium aluminum hydride offers the advantage of simultaneously removing water and peroxides. If the olefin requires a large amount of LAH for purification (use test for excess hydride), the material should be filtered prior to distillation to lessen the possibility of an exothermic polymerization reaction catalyzed by aluminum salts. If the olefin is suspected of containing large amounts of peroxides, it should be tested and the peroxides removed by standard chemical means prior to distillation with LAH (Ref. 44c, p. 437). Olefins boiling above 100° should be distilled from LAH under reduced pressure, and the distillation *never* allowed to go to dryness. Since vacuum distillations of olefins from active hydrides are often plagued by foaming, a splash guard or a long Vigreaux column should be employed. Refer to the discussion elsewhere in this section for other precautions in using lithium aluminum hydride.

Functionally substituted olefins may be dried by appropriate drying agents (Molecular Sieves, Drierite, magnesium sulfate). High boiling materials can often be adequately dried by distillation. If peroxidic impurities are suspected, they should be removed by washing with concentrated aqueous ferrous sulfate in hydrochloric acid or other standard techniques prior to drying.[44c]

Solubilities of Hydrides and Reagents

The solvents, diethyl ether, tetrahydrofuran, tetrahydropyran, monoglyme, and diglyme, are especially valuable for the reactions involving borohydrides and boranes described in this volume. A summary of pertinent solubility data for these hydrides and related reagents is contained in **T9.4**.

Table 9.4. Solubilities of Hydrides and Reagents in Ethers at 25° in g/100 g Solvent

Compound	Diethyl Ether	Tetrahydro-furan	Tetrahydro-pyran	Monoglyme	Diglyme
LiF	10^{-3}	10^{-3}	10^{-4}	10^{-4}	10^{-4}
LiCl	2×10^{-3}	4.19	0.56	0.45	1.36
LiBr	12.85	38.64	24.06	17.00	27.45
$LiBF_4$	1.90	71	Sol.	Sol.	Sol.
$LiBH_4$	4.28	28	Sol.	12.1	16.2
$LiB(OCH_3)_4$	10^{-2}	10^{-2}	–	10^{-2}	Insol.
$NaBH_4$	Insol.	0.1	–	0.8	6.6
$NaB(OCH_3)_4$	Insol.	94.5	69	2.7	10^{-4}
KBH_4	Insol.	0.23	–	0.0	0.0
B_2H_6	1.1[a]	8.1[a]	–	0.66[b]	0.76[b]
$LiAlH_4$	35.40[c]	13	–	10	3.6[d]

[a] At 20° and 1 atm.
[b] At room temperature and 0.9 atm pressure.
[c] To get this concentration, a previously prepared solution must be concentrated.
[d] 0.95*M* at 25°C.

Preparation of a Tetrahydrofuran Solution of Lithium Aluminum Hydride[51]

Using a 1-ℓ flask, the apparatus shown in **F9.5** is oven-dried, assembled hot, and allowed to cool under a stream of nitrogen. Dry THF, 800 ml, is added through the septum inlet. With nitrogen flowing slowly in through the septum inlet, 50 g of 95+% purity LAH (1.25 moles) is added to the flask by temporarily removing the connecting tube **(N1, N2)**. The mixture is stirred at least 2 hours. A 1-ℓ flask and a 500-ml filter chamber **(F9.27)**, charged with sufficient Celite to form a 2-in. filter cake, is oven dried, assembled hot, and allowed to cool under a stream of nitrogen. The crude solution of LAH is transferred into the filter via double-ended needle through the septum inlet. A small positive pressure of nitrogen facilitates the slow filtration **(S9.6)**. The filtrate is a crystal clear solution about 1.5*M* in LAH. The exact concentration of the solution can be determined by

[51] H. C. Brown and P. M. Weissman, *J. Amer. Chem. Soc.*, 87, 5614 (1965).

hydrolysis of an aliquot and measurement of the amount of hydrogen evolved **(S9.10)**.

Notes

1. Solid LAH is not spontaneously flammable, but it may ignite on rubbing or vigorous grinding. Contact with liquid water may cause ignition. The solid is best handled in a glove bag to prevent decomposition by atmospheric moisture. Solutions of LAH should be handled under nitrogen. Spillage may spontaneously ignite after a brief exposure to air. MetL-X, dry sand, or calcium carbonate (powdered limestone) may be used to control LAH fires. *Never use water, soda acid, chemical or carbon dioxide fire extinguishers on LAH fires.* Care should be exercised in avoiding skin contact or breathing the dust. Waste LAH solutions are best destroyed by the dropwise addition of ethyl acetate.[48]
2. The LAH should be added to the THF and *not* vice versa.

Purification of Boron Trifluoride Diethyl Etherate[50]

Commercial BF_3:Et_2O is suitable for most purposes **(N1)**. The material may be purified by distillation under an inert atmosphere **(S9.7)**. Approximately 500 ml of commercial BF_3:Et_2O, 2 g of CaH_2 **(N1)**, and 10 ml of Et_2O are added to a 1-ℓ pot. The mixture is distilled under vacuum. A moderate forerun (30 ml) is taken, and the pure BF_3:Et_2O is distilled at 60° and 20 mm **(N2)**.

Notes

1. Commercial BF_3:Et_2O may contain certain acidic impurities, such as arsenic and phosphorus compounds, which are reduced to foul smelling hydrides, PH_3 and AsH_3, during the preparation of BH_3 from undistilled BF_3:Et_2O. Calcium hydride serves to remove certain of these impurities–it also reduces bumping during the distillation.
2. BF_3:Et_2O may darken on standing and should be protected from air and light.

Preparation of Boron Trifluoride Tetrahydrofuranate

Using a 1-ℓ flask, the apparatus shown in **F9.5** is oven-dried, assembled hot, and allowed to cool under a stream of nitrogen. Distilled BF_3:Et_2O, 248 g (221 ml, 1.75 moles), is added to the flask through the septum inlet. Stirring is begun and the flask cooled in an ice-water bath. Dry THF, 126 g (142 ml, 1.75 moles), is added slowly through the septum inlet. Diethyl ether is removed by water aspirator until the flask reaches constant weight. The resultant BF_3:THF may be used without further purification **(N1)**.

Notes

1. BF_3:THF darkens on standing and should be protected from air and light. The product can be distilled under reduced pressure, bp 68-69° at 3 mm, but undergoes some polymerization on distillation or standing. Consequently, it is best to use freshly prepared material.

Preparation of Boron Trifluoride in Diglyme[52]

Using a 200-ml flask, the apparatus shown in **F9.5** is oven-dried, assembled hot, and allowed to cool under a stream of nitrogen. Dry, deoxygenated diglyme, 50 ml, is added to the flask through the septum inlet. Stirring is begun, and the flask is cooled in an ice-water bath. Distilled BF_3:Et_2O, 25 ml, is added slowly through the septum inlet. Diethyl ether is removed by vacuum, (~5 mm) for 20 min maintaining the flask at 20-25°. The resulting product, 3.65*M* in BF_3, is used without further purification. Notice that the product is a solution of boron trifluoride diglymate in diglyme. The material darkens rapidly and should be utilized soon after preparation.

Purification of Sodium Borohydride[53]

Using a 500-ml flask, the apparatus shown in **F9.5** is oven-dried, assembled hot, and allowed to cool under a stream of nitrogen. The flask is charged with 25 g of $NaBH_4$. Dry, deoxygenated diglyme, 300 ml, is added through the septum inlet. The mixture is heated to 50° with stirring for 30 minutes to dissolve most of the solid. The hot solution is transferred and filtered through a double-ended needle fitted with an in-line filter (**F9.26**) into a dried, flushed-out flask (**S9.6**). The filtered solution is allowed to cool slowly to room temperature, and then the flask is cooled with an ice-water bath. When crystallization is complete, the supernatant liquid is decanted with the aid of a double-ended needle (**S9.6**). The crystalline complex, $NaBH_4$·diglyme, is heated to 60° for 4 hours under vacuum to remove the diglyme. The resulting product is greater than 99% pure (**N1**).

Notes

1. Sodium borohydride and its solutions are relatively safe materials to handle. The dry solid is shock stable, but flammable. The solid is hygroscopic and picks up water to form the dihydrate which undergoes slow hydrolysis. Consequently, the solid should be stored in nitrogen-flushed sealed con-

[52] H. C. Brown and G. Zweifel, *J. Amer. Chem. Soc.*, 83, 1241 (1961).
[53] H. C. Brown, E. J. Mead, and B. C. Subba Rao, *J. Amer. Chem. Soc.*, 77, 6209 (1955).

tainers. Sodium borohydride solutions in diglyme are stable. Stable solutions in isopropyl alcohol (0.1*M*) can be prepared. Aqueous solutions can be stabilized by the presence of alkali. If sodium borohydride solutions are exposed to dilute acids, finely divided metals, or certain metal salts, large quantities of hydrogen may be generated rapidly. Contact of sodium borohydride solid or solutions with concentrated acids or strong Lewis acids can result in the formation of diborane gas or spontaneously inflammable metal borohydrides (such as aluminum borohydride). MetL-X may be used to control borohydride fires. Solutions of sodium borohydride are basic and skin contact can cause caustic burns.

Purification of Lithium Borohydride

Using a 500-ml flask, the apparatus shown in **F9.5** is oven-dried, assembled hot, and allowed to cool under a stream of nitrogen. The flask is charged with 5 g of $LiBH_4$ (**N1**). Anhydrous diethyl ether, 120 ml, is added through the septum inlet. The mixture is stirred until most of the solid dissolves. The resulting solution is transferred and filtered through a double-ended needle fitted with an inline filter (**F9.26**) into a dried, flushed out flask (**S9.6**). The ether is removed under vacuum to give first the monoetherate, $LiBH_4 \cdot OEt_2$, and then the hemietherate, $LiBH_4 \cdot 1/2Et_2O$ (**N2**). On further treatment, unsolvated lithium borohydride is obtained (**N3**).

Notes

1. Lithium borohydride is best handled in a glove bag to prevent atmospheric decomposition. Caution should be observed to avoid breathing the caustic and irritating dust. Lithium borohydride is flammable, hygroscopic, and sometimes ignites on contact with water. Contact with cellulosic materials may cause spontaneous combustion. A white oxide coating forms on the surface after about 2-3 minutes exposure to air. MetL-X, dry sand, or calcium carbonate may be used to control lithium borohydride fires.
2. The monoetherate exhibits a dissociation pressure of 90 mm at 25° and passes through the hemietherate with a dissociation pressure of 46 mm at 25°.
3. Purified material must be stored in nitrogen-flushed sealed containers to avoid decomposition.

APPENDIX

SOURCES OF TECHNICAL LITERATURE

Reagents

Literature concerning the chemical and physical properties of organometallic reagents as well as recommended safety and handling procedures may be obtained from the following companies:

Aldrich Chemical Company
940 West St. Paul Avenue
Milwaukee, Wisconsin 53233

Arapahoe Chemicals
2855 Walnut Street
P.O. Box 511
Boulder, Colorado 80302

Callery Chemical Company
Callery, Pennsylvania 16024

Ethyl Corporation
Industrial Chemical Division
Ethyl Tower, 451 Florida
Baton Rouge, Louisiana 70801

Foote Mineral Company
Chemical Products
Chemical & Minerals Division
Route 100
Exton, Pennsylvania 19341

Lithium Corporation of America
Technical Service Division
Bessemer City, North Carolina 28016

Stauffer Chemical Company
Specialty Chemical Division
10 South Riverside Plaza
Chicago, Illinois 60606

Texas Alkyls Inc.
6910 Fannin Street
Houston, Texas 77025

Ventron/Alfa
Metal Chemicals Division
Congress Street
Beverly, Massachusetts 01915

Equipment

Information concerning septa and elastomers may be obtained from:

Parker Seal Company
10567 Jefferson Boulevard
Culver City, California 90230

The West Company
Phoenixville, Pennsylvania 19460

Pierce Chemical Company
P.O. Box 117
Rockford, Illinois 61105

Information regarding ampules, hypodermic syringes and acessories may be obtained from:

Ace Glass Company
1430 Northwest Boulevard
P. O. Box 688
Vineland, New Jersy 08360

Becton-Dickinson & Co.
Rutherford, New Jersey 07070

Hamilton Company
4960 Energy Way
P. O. Box 7500
Reno, Nevada 89502

Popper & Sons, Inc.
300 Denton Avenue
New Hyde Park, New York 11040

Wheaton Scientific
1000 N. Tenth Street
Millville, New Jersey 08332

Sage Instruments
Division of Orion Research Inc.
11 Blackstone St.
Cambridge, Massachusetts 02139

Information about BTS catalyst may be obtained trom:

Ace Glass Company
1430 Northwest Boulevard
P. O. Box 688
Vineland, New Jersey 08360

BASF Colors & Chemicals, Inc.
845 Third Avenue
New York, New York 10022

INDEX